THE RIVER AND THE ROCK
RIVER POTHOLES OF WALES
DEWI ROBERTS • STEPHEN TOOTH • HYWEL GRIFFITHS

First published: 2022
© text and images:: Dewi Roberts, Hywel Griffiths, Stephen Tooth
Other contributions are acknowledged where appropriate in the text.

All rights reserved.
No part of this publication
may be reproduced, stored in a retrieval system,
or transmitted in any form or by any means, electronic,
electrostatic, magnetic tape, mechanical, photocopying,
recording, or otherwise, without prior permission
of the publishers, Gwasg Carreg Gwalch,
12 Iard yr Orsaf, Llanrwst, Dyffryn Conwy, Cymru LL26 0EH.

ISBN: 978-1-84527-887-8

Published with the financial support of the Books Council of Wales

Cover design and layout: Eleri Owen

Published by Gwasg Carreg Gwalch,
12 Iard yr Orsaf, Llanrwst, Dyffryn Conwy, Cymru LL26 0EH.
tel: 01492 642031
fax: 01492 642502
email: books@carreg-gwalch.cymru
website: www.carreg-gwalch.cymru

Printed and published in Wales.

CONTENT

Introduction 5

Section 1 The Geomorphology of Potholes 7

Section 2 The Ecology of Potholes 27

Section 3 Potholes and History 41

Section 4 Potholes and the Creative Arts 55

Section 5 International Examples of Potholes 67

Section 6 Potholes and Health 81

Section 7 Visiting Welsh Potholes 91

Section 8 Glossary 107

Section 9 Further Reading and Online Resources 113

Section 10 Suggested Educational Activities 119

Acknowledgements and Biographies 126

INTRODUCTION

Wales is a land of mountains, hills, valleys and coastal lowlands that is drained by networks of stunning rivers. Rivers, their valleys and their catchments often define their areas geographically and culturally, with examples including the Banwy valley ('Dyffryn Banwy') and Conwy valley ('Dyffryn Conwy'). Particularly in the Welsh uplands, a combination of high precipitation (rain, snow), steep terrain, and outcrops of hard bedrock means that waterfalls, rapids and gorges are some of the most prominent river features. Many of these features are well known, commonly form major tourist attractions, and may be within national parks or other protected areas. Yet along these rivers, there are many less commonly appreciated but equally fascinating features. In this book, we focus on river potholes: roughly circular depressions eroded into the rock forming channel beds and banks. Potholes and an array of related natural sculptural forms are found mainly in the upper sections of Welsh rivers formed in hard bedrock but sometimes also in lowland sections formed in softer sediment such as clay. Our aim is to raise awareness of the significance of potholes for river landscape development, ecology, and society, so that these too may be seen as an important part of our natural and cultural heritage, worthy of our respect and protection.

We start by looking at the science of potholes and provide an overview of the geomorphological processes involved in their formation and development, and their links to features such as waterfalls, rapids and gorges (Section 1). We then address the ecology of potholes by discussing how a range of wildlife makes use of them, and some of the human impacts that disrupt the natural ecology (Section 2). Rivers in general are a major landscape attraction for many people and we outline some of the history, stories and legends associated with potholes (Section 3), and provide examples of some of the ways artists and writers have responded to potholes and related features (Section 4). Looking beyond Wales, we provide examples of potholes in other rivers worldwide to show how these features also have natural and cultural significance (Section 5). We also discuss how potholes can be important for people's physical and mental wellbeing (Section 6).

The remaining sections then provide additional information and resources. We list examples of locations in Wales where potholes can be seen, and give suggestions for possible itineraries (Section 7). To assist with pothole and river terminology that may be unfamiliar to readers, we provide a glossary (Section 8). We also identify further reading and online resources (Section 9) and provide some suggested educational activities based around potholes and related features (Section 10).

This book is not a definitive guide to river potholes, but rather one that provides a starting point for a glimpse into their fascinating worlds. We like to view potholes and related features as natural sculptures, artwork that is continually being created by the action of river flow and sediment passing over rock, and we hope that others may come to share and develop this perspective.

Afon Elan near Pont ar Elan, Powys (HG).

SECTION 1

• THE GEOMORPHOLOGY OF POTHOLES •

River systems essentially are the 'arteries' of the land, transporting water, sediment and nutrients from upland to lowland areas. By doing so, rivers shape the land, sustain ecological systems, and support a wide range of human activities.

These simple statements hide the fact that river system characteristics can vary greatly. A basic distinction is commonly drawn between rivers that flow mainly over solid rock ('bedrock rivers') and those that flow through mud, sand or gravel sediments ('alluvial rivers'). Traditionally, scientists who specialise in the study of river processes and forms (fluvial geomorphologists) have tended to regard bedrock and alluvial rivers as being fundamentally different in character, with bedrock rivers being shaped primarily by rock hardness, and alluvial rivers being shaped primarily by flow and sediment transport processes.

In reality, however, this distinction is not so straightforward. Worldwide, many rivers have characteristics of both bedrock and alluvial rivers. In these rivers – perhaps best termed 'mixed bedrock-alluvial' rivers – certain reaches may be characterised by sand and gravel sediment that forms only a thin or patchy veener over rock, so that rock essentially forms the channel bed or banks. In these types of rivers, the combination of rock hardness, flow, and sediment transport processes can create intricate and beautiful natural sculptures. Some of the most striking sculptural forms are river potholes, which are roughly circular depressions eroded into the rock forming the channel bed and banks. In some locations, potholes can also be eroded into large, immobile boulders that have fallen from steep hillsides into the channel. By contributing to the progressive wearing away of rock, potholes can play a key role in the creation of some of the world's most treasured river scenery, including waterfalls, rapids and gorges, but also help create ecological habitats, inspire artists and writers, and provide settings for myth and legend. This river scenery can draw tourists from far and wide, and create environments where our physical and mental wellbeing can be enhanced.

Many Welsh rivers provide great examples of potholes and related river scenery. Extensive upland areas such as Eryri/Snowdonia, the Berwyn Range, the Cambrian Mountains, the Brecon Beacons and the Preseli Hills are drained by rivers that flow through deep, steeply sloping valleys and descend rapidly to relatively narrow coastal lowlands (Figure 1.1). Many rivers tend to be bedrock or mixed bedrock-alluvial in their upper reaches, perhaps changing to a more alluvial character farther downstream. Other rivers alternate along their length from bedrock or mixed bedrock-alluvial to alluvial and back again, perhaps repeatedly in their journey from their headwaters to the coast. Consequently, spectacular examples of potholes, waterfalls, rapids and gorges can be found along many Welsh rivers (Figures 1.2 and 1.3), but potholes in particular remain little known and often overlooked elements of the landscape, with their ecological and wider cultural significance commonly underappreciated.

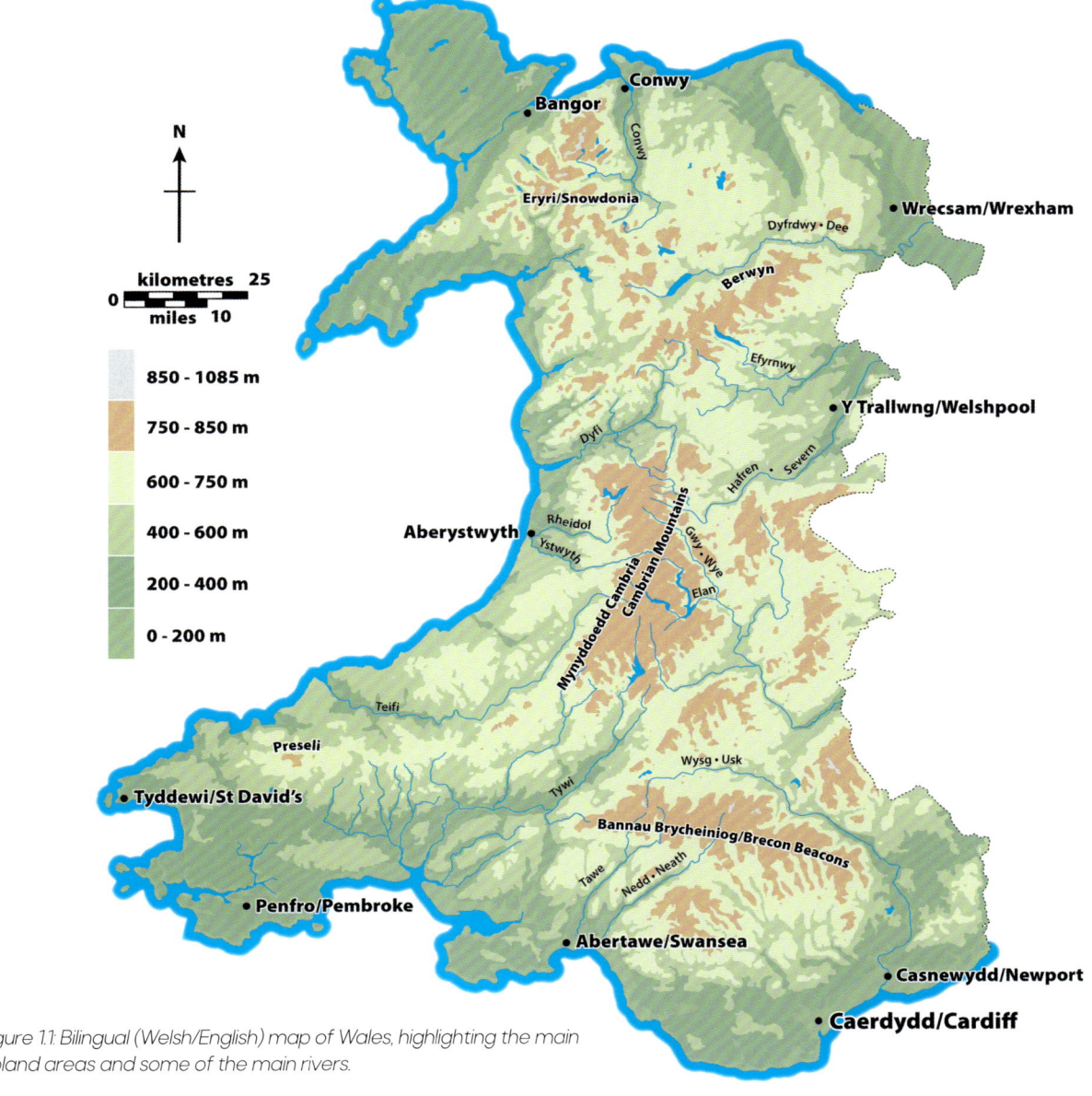

Figure 1.1: Bilingual (Welsh/English) map of Wales, highlighting the main upland areas and some of the main rivers.

Figure 1.2: The Afon Irfon at Camddwr Bleiddiad, Abergwesyn, Powys (DR).

Figure 1.3: Sgwd Ddwli, Afon Nedd Fechan, Powys (DR).

The purpose of this section of the book is to outline some aspects of the geomorphology of potholes. Drawing on examples from Wales and farther afield, we ask a series of questions. What exactly are river potholes, and are there other kinds of potholes? How do river potholes form and develop over time? How fast do potholes develop and how old can they be? How is pothole development linked with the formation of waterfalls, rapids and gorges? Some of the answers to these questions involve a little technical information, which we provide through some conceptual diagrams and descriptions. While it is certainly not necessary to understand everything to enjoy subsequent sections, we believe that the answers may help to enhance appreciation of the ecological and wider cultural significance of potholes.

WHAT ARE RIVER POTHOLES?

Despite the common occurrence of potholes in many bedrock and mixed-bedrock alluvial rivers worldwide, there is no universally agreed definition of the term 'pothole'. Nevertheless, while the definitions provided in scientific publications differ in detail, they all tend to stress the main characteristics of potholes, including their common association with turbulent river flow, their erosion into bedrock, and their roughly circular shapes.

While the term 'pothole' may be familiar to some readers, and form part of everyday language for describing river features, a variety of other English-language terms have also been applied to potholes and pothole-like features, both within the UK and overseas. Examples include 'rock mill', 'churn hole', 'eddy mill', 'scour hole', 'rumbling hole', 'swirlhole', 'kettle', and 'tolmen' (possibly specifically in Cornwall). Languages other than English commonly do not have a literal equivalent of 'pothole' and so introduce an even wider variety of terms. In Welsh, the terms 'ceudwll' (enclosed hole), 'ceubwll' (enclosed pool), or 'trodwll' (turn hole) are commonly used. In Afrikaans, the term 'kolkgat' (plural 'kolkgate') is commonly used to refer to river potholes. In scientific writing, 'kolk' is used in a technical sense to refer to an upward-rising turbulent vortex in water. Used in conjunction with 'gat' ('hole'), this term draws attention to the turbulent vortices that are involved in pothole formation (see 'How do river potholes form and develop?' and also Section 5).

Confusion can arise from the fact that other natural bedrock features unrelated to river action have also been termed potholes. These include cylindrical depressions found on bedrock in coastal locations subject to wave and tidal action ('marine potholes') as well as depressions found on slow-moving or surging icy glacier surfaces subject to melting and to pressure and tensional forces. In North America, periodically inundated depressions in formerly glaciated, hummocky terrain are known as 'prairie potholes'. The term 'pothole' is also used in relation to various depressions etched into bedrock but that are located far away from the influence of river, marine or glacial influences, and instead arise from natural subaerial weathering processes (i.e. the action of rain, sun, frost and wind on exposed bedrock surfaces). These weathering depressions are widely referred to as 'potholes' in parts of the southwest United States but elsewhere in the United States and in other countries, the same features fall under a variety of other terms including 'tinajas', 'gnammas', 'rock basins', and 'solutional pits'.

Natural depressions aside, the term 'pothole' or variants are also used widely for a variety of other features and activities. In the UK and some other countries, 'potholing' is a leisure activity and sport that involves exploration of subterranean vertical cave systems ('cave potholes'), while in many parts of the English-speaking world, depressions in dirt or tarmacked road surfaces are also referred to as potholes. To distinguish from river potholes ('kolkgate'), Afrikaans has a different word for dangerous road potholes, namely 'slaggate' (literally translated as 'impact hole' or 'slaughter hole'). In this book, we use the term 'pothole' as shorthand for 'river pothole'.

HOW DO RIVER POTHOLES FORM AND DEVELOP?

Questions regarding the formation and development of potholes have interested scientists for a century or more. In rare cases, potholes can form in soft but cohesive clay but are more common on a range of hard rock types, including igneous (e.g. basalt, granite), metamorphic (e.g. quartzite, gneiss) and sedimentary (e.g. sandstone, shale) rocks. There is general agreement that pothole formation is related to the grinding and polishing of this bedrock by sediment particles (mainly sand and gravel) that are swirled around in initially small depressions by turbulent eddies in fast-flowing river water (Figure 1.4). Potholes range from very small (centimetre scale) to very large (many metres scale), which suggests that initially small potholes may grow in size over time, eventually forming larger potholes.

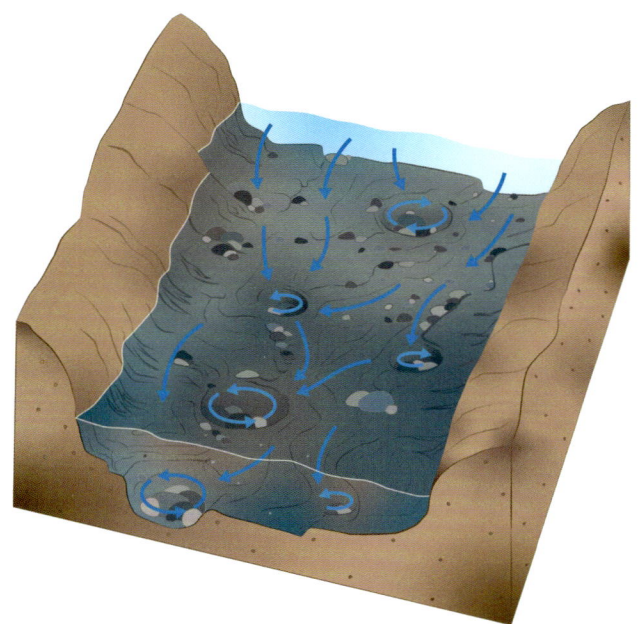

Figure 1.4: Sketch diagram showing potholes of various shapes and sizes on a river bed subject to turbulent flow.

Despite this general agreement, pothole formation and development continue to be active topics for scholarly enquiry, and new data, information and insights continue to be generated. In general terms, however, potholes can be considered as undergoing a 'life cycle' that involves three main phases. We can call these phases birth, growth and decay.

1. Birth

The birth phase of potholes refers to the establishment of initial depressions on bedrock surfaces or immobile boulders that are subject to the influence of river flow. Irregularities in otherwise initially near-smooth bedrock or boulder surfaces can arise from cracks, fractures or joints (Figure 1.5), from foliation (preferred orientation) of the minerals comprising igneous or metamorphic rocks, or from the layering in sedimentary rocks. Uneven erosion can occur in fast-flowing water, resulting in undulations in bedrock surfaces (Figure 1.5), perhaps also with small blocks of bedrock being removed owing to pressure fluctuations in overpassing flow or small pits being generated by the impact of transported gravel. The initial depressions thus may be of various shapes and sizes but, once formed, provide areas within which turbulent eddies can be generated, and transported sediment particles can be trapped, at least temporarily. Over time, the swirling action of water and sediment within these initial depressions leads to additional bedrock erosion, especially where the trapped sediment particles are of a harder rock type than the rock in which the pothole is forming. In many Welsh rivers, for instance, quartz particles transported from upstream outcrops are harder than the rock types forming the river bed and banks farther downstream, and so can be potent erosional agents. In effect, rivers are using water and sediment as tools to widen and deepen the initial depression and essentially are drilling holes into solid bedrock. Initially, the enlarging depression may have a roughly hemispherical (i.e. semi-circular) shape but if, over time, it assumes a more cylindrical outline, a pothole can be said to have been born (Figure 1.6).

Figure 1.6: A typical pothole developed in granitic rocks on the Orange River near Augrabies Falls, western South Africa. Many potholes start as small, hemispherical features (left and right of image) but as they enlarge they develop more cylindrical forms (centre foreground) or may evolve into other sculpted forms such as furrows. Sediment enters and exits the pothole in highly turbulent flood flows (ST).

Figure 1.5: Irregularities in a flood-eroded, granitic bedrock surface that may be the initial starting point for potholes: Vaal River near Parys, South Africa. Flood flow direction was from left to right (ST).

2. Growth

The growth phase of potholes refers to the development of the more cylindrical depressions truly deserving of the term 'pothole' (Figure 1.6). For the pothole to develop, the rate of deepening must equal or exceed the rate of erosion on the bedrock surface surrounding the pothole. If not, then the initial pothole will become shallower over time or may be entirely removed by erosion (Figure 1.7). Those potholes that do deepen faster than the rate of erosion on the surrounding bedrock will survive and grow over time, both by increasing depth and width. Measured data from potholes in Wales and overseas show that many potholes tend to deepen at a roughly similar rate to their widening, and this provides important insights into the processes responsible for pothole development.

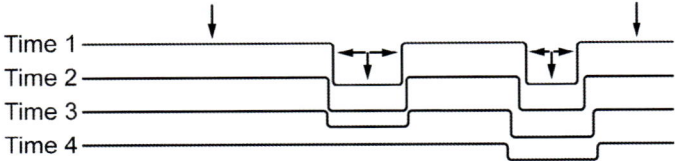

direction of channel bed erosion/pothole growth

Figure 1.7: Conceptual diagram (above) illustrating pothole growth dynamics in cases where channel bed erosion on the surfaces surrounding the potholes is occurring at a faster rate than the pothole is deepening. Over time, this leads to pothole loss (left pathway) or to pothole shallowing (right pathway). Example of a pothole eroded into granitic rocks on the Orange River near Augrabies Falls, western South Africa (below). While the pothole appears to be actively growing, erosion of the rock surrounding the pothole is also clearly occurring, so over time this pothole may become shallower or be lost (ST).

Hydraulic studies suggest that flow entering a pothole initially spirals around the walls of the pothole, before rising up out of the centre of the pothole as a coherent eddy or 'kolk' (see definition above). But the sediment that is transported by the flow is the key factor. To enable pothole deepening, bedrock has to be eroded from the floor of the pothole (see shaded area in Figure 1.8). This implies that the circulation of larger sediment particles around the pothole floor (bed load sediment) is an important agent of erosion (Figures 1.2, 1.6 and 1.7). These larger particles are sometimes referred to as 'grinders'. But to enable continued widening, more and more bedrock has to be eroded from the enlarging area of the pothole walls (see unshaded area in Figure 1.8). This implies that other important agents of erosion include the sediment that enters the potholes and spirals around the pothole walls to the pothole floor, as well as the sediment that may be kept aloft in the spiralling flow at the margins of the pothole (saltating or suspended sediment). These spiralling particles from time to time impact on the walls of the potholes, over time leading to a 'polishing' or 'sandpapering' effect that slowly and incrementally removes bedrock. The smooth-sided walls and floors of many potholes attest to the effectiveness of this process.

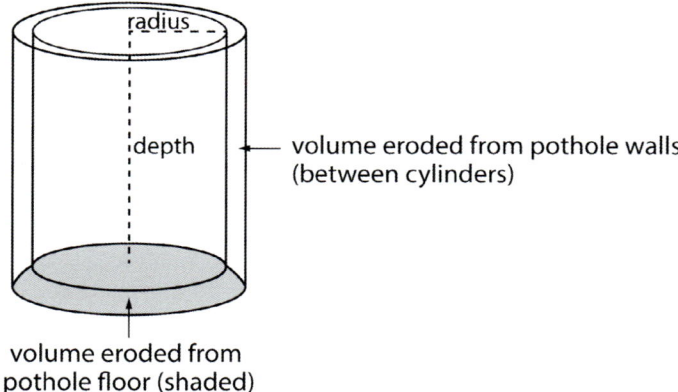

Figure 1.8: Schematic diagram of pothole growth, showing an initial cylindrical form that enlarges over time. Simultaneous erosion of the pothole walls and floors causes both pothole radius (hence, width) and depth to increase. The 3D geometry of the cylindrical pothole means that the increasing width contributes proportionally more to the volume of rock removed by erosion (Source: simplified from Springer, G.S, Tooth, S. and Wohl, E.E. (2005). Dynamics of pothole growth as defined by field data and geometrical description. Journal of Geophysical Research: Earth Surface, 110: F04010).

These descriptions of sediment movement in potholes simplify what in reality are extremely complex patterns. For instance, field experiments in Welsh potholes using deliberately introduced pebbles ('seeded' clasts) show complex patterns of sediment exchange in turbulent flows, with pebbles moving from potholes to river beds and vice versa (Figure 1.9).

Complications to the idealised pothole growth dynamic shown in Figure 1.8 can arise. River potholes are rarely perfectly cylindrical, neither along Welsh rivers nor in other settings. Researchers have long recognised that deviations from a near-cylindrical shape may arise from

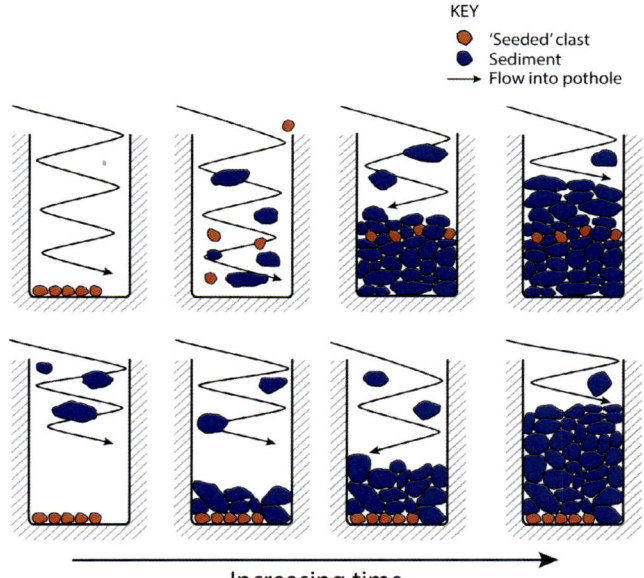

Figure 1.9: Conceptual diagram indicating how sediment particles may behave under the influence of turbulent flow vortices in a pothole. For simplicity, flow is only shown entering the pothole but in reality also spirals out of the pothole. In the top scenario, the incoming vortex reaches the floor of the pothole, entrains the 'seeded' clasts, and over time some are lost. Other sediment eventually comes into the pothole, burying the remaining 'seeded' clasts partway up the infill. In the bottom scenario, the vortex does not reach the floor of the pothole and so the 'seeded' clasts are not entrained. As other sediment comes into the potholes, the 'seeded' clasts are buried (Source: adapted from Richardson, J. (2013). Controls on the location, development and significance of bedrock reaches on the middle River Rheidol, west Wales. Unpublished MPhil thesis, Aberystwyth University, 306 pp.).

various factors, including rock properties, the complexities of river flow and sediment transport, other bedrock erosional processes, and interactions between potholes.

For instance, intrinsic rock properties, such as mineral

variations within igneous rocks, the foliation within metamorphic rocks, or the layers within sedimentary rocks may lead to small-scale variations in bedrock erosional resistance. Cracks, fractures or joints may also provide variations in erosional resistance. In many Welsh bedrock and mixed bedrock-alluvial rivers, just as in such rivers elsewhere, many potholes show clear evidence of preferential widening and/or deepening along the less resistant parts of the rock outcrop, commonly leading to oval, elliptical or more distorted forms (Figure 1.10)

Other deviations from near-cylindrical shapes may also result from the complexities of river flow and sediment transport. In turbulent river flow, transported particles may be continually interchanging between the bed load and the suspended sediment fraction (Figure 1.9). Larger particles may enter the pothole in suspension, perhaps impacting on the pothole walls and contributing to wall erosion and pothole widening as they spiral downwards,

Figure 1.10: Examples of oval and elongated potholes on the upper Afon Tywi, on the Powys/Ceredigion border. Flood flow is from left to right and the potholes can be seen to have grown preferentially in a direction that is obliquely downstream (DR).

but then may continue to spiral around the floor of the pothole and contribute to floor erosion and pothole deepening. But the situation is complex, as too much sediment on the floor of the pothole may end up preventing sediment circulation and actually bury the bedrock (Figures 1.9 and 1.11), effectively protecting the floor from the impacts of other particles and leading to a temporary pause in pothole deepening. In such a situation, erosion will be focused on the pothole walls at some distance above the floor of the pothole. If sediment is later removed from the floor of the pothole (e.g. under more vigorous vortices that may develop during larger floods), this may reveal more irregular walls (Figure 1.12).

Figure 1.11: Pothole eroded into granitic rocks on the Sabie River in the Kruger National Park, eastern South Africa. The pothole has been nearly filled to the rim with pebbles and cobbles. The deeper sediment may remain largely immobile, even in large floods, and so is effectively protecting the pothole floor from erosion (ST).

Figure 1.12: Pothole eroded into granitic rocks on the Vaal River, South Africa. Sediment within the pothole has been removed during a recent flood, revealing irregular walls and a width that decreases with depth (ST).

Other processes of bedrock erosion may also lead to deviations from a near-cylindrical shape. For the most part, potholes develop as result of bedrock abrasion, the grain-by-grain removal from bedrock surfaces resulting from the impacts (grinding, polishing) of bed load, saltating

Figure 1.13: Example of a pothole eroded into quartzite rocks on the Vaal River, South Africa. The pothole was initially about 30 cm deep but removal of joint- or fracture-bounded blocks of rock from part of the pothole walls has reduced its depth by about half (ST).

Figure 1.14: Example of potholes eroded into quartzite rocks on the Vaal River, South Africa, that have coalesced through removal of joint- or fracture-bounded blocks of rock. Water is retained within the deeper pothole (lower right). A small, secondary pothole has formed in the floor of the shallower pothole (centre) (ST).

Figure 1.15: Example of coalesced potholes beneath the waters of the Afon Efyrnwy/Vyrnwy, Powys (DR).

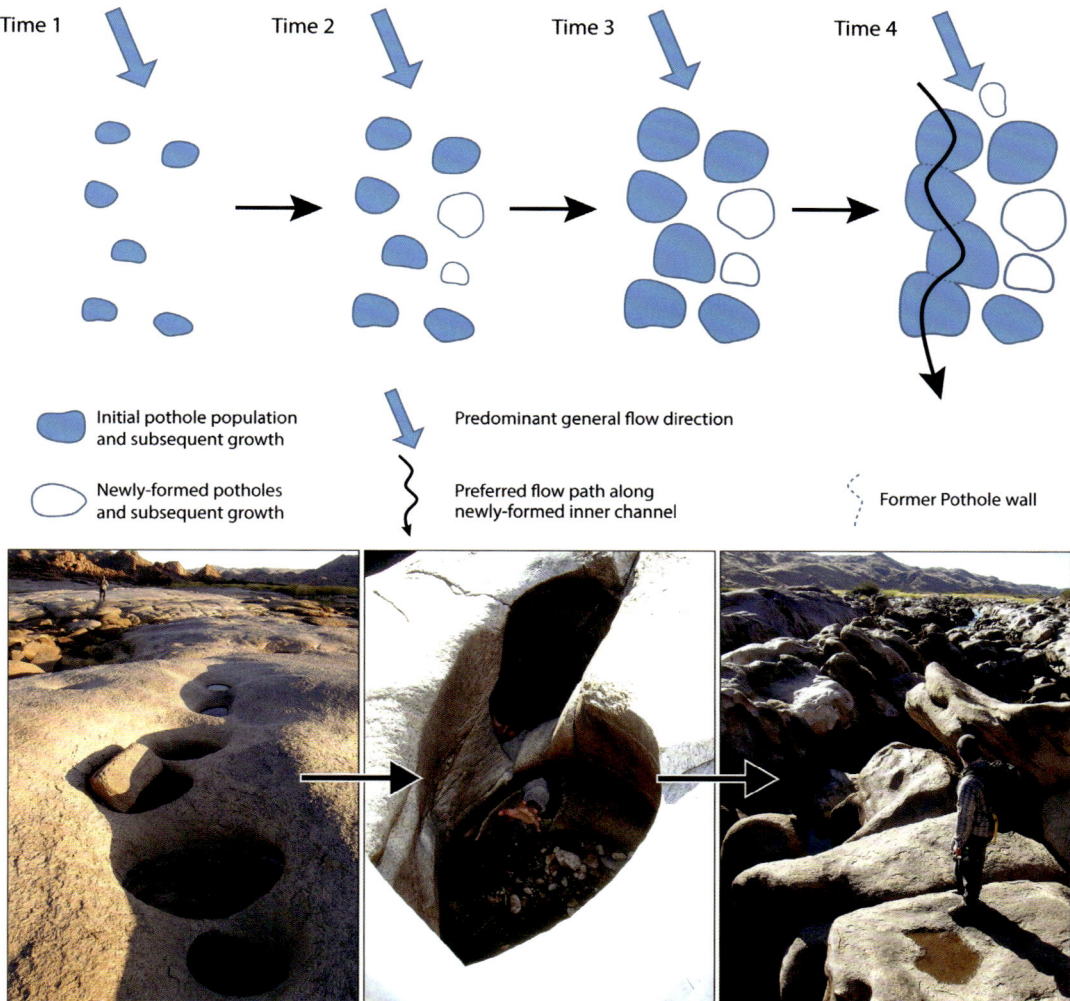

Figure 1.16: Conceptual diagram (above) and photographs (below) showing how, over time, pothole growth can lead to pothole coalescence, and ultimately to formation of a preferred flow path along an 'inner channel' carved through resistant rock. The photographs are from the lower Orange River, eroded into granitic rocks on the South African/Namibian border (local flow directions are away from the camera: left) closely spaced but individual potholes varying in width and depth; middle) individual deeper, wider potholes that have started to coalesce through erosion of the potholes walls; right) extensive coalesce of numerous potholes, such that individual potholes are hard to distinguish (ST).

or suspended sediment. In some cases, however, individual joint- or fracture-bounded blocks of rock may be eroded from pothole walls or surrounding surfaces as a result of bedrock plucking ('hydraulic quarrying') during flood flows (Figure 1.13). Plucking can destroy incipient or small potholes entirely but around larger potholes can lead to removal of the upper part of pothole walls, and thus to instant decreases in pothole depth, something that might be termed 'arrested growth'. In other situations, removal of only a portion of the upper pothole walls may occur, with the consequence that many potholes have rims that vary in elevation around their circumference (Figure 1.13). By removing parts of pothole walls, plucking may also facilitate the merging (coalescence) of two neighbouring potholes, leading to a complex, compound pothole that deviates further from the idealised cylindrical shape (Figures 1.14 and 1.15).

3. Decay

The decay phase of a pothole refers to the loss of individual pothole identity. As outlined above, plucking can contribute to pothole loss or shape distortion during the birth or growth phases, but even those potholes that can be said to have reached maturity ultimately may undergo partial or complete decay. Over time, individual potholes grow in size, both by deepening and widening, and in some cases coalesce with neighbouring potholes to form larger, compound potholes (Figures 1.14 and 1.15). In the early stages of coalescence, pothole identity may be partly maintained but further coalescence with additional potholes may lead to the loss of large parts of the pothole walls and effectively loss of distinctive pothole features (Figure 1.16).

HOW FAST DO POTHOLES DEVELOP AND HOW OLD CAN THEY BE?

In most locations worldwide, the rate of pothole development and the age of existing potholes is not well known. In a few settings, very rapid development appears to have occurred. For instance, in the Channeled Scablands of the northwestern United States, very large potholes up to 30 m wide and 5 m deep formed in catastrophic floods during the last Ice Age (see Section 5). On more recent timescales, rapid pothole development has also been documented. Along the Upper Ukak River, Alaska, potholes 4-6 m wide and 2-3 m deep developed rapidly in sedimentary rocks (sandstones, siltstones) following a 1912 volcanic eruption that displaced the original river course, while along the Indrayani River, India, potholes up to about 1 m wide and 1.3 m deep formed in human-made channels carved in basalt over about 60 years (see further reading in Section 9). In these instances, therefore, potholes have either formed essentially instantaneously, or average growth rates of pothole depths and widths have been greater than a few millimetres a year.

In most other rivers in Wales and farther afield, however, similarly rapid pothole development is unlikely. Potholes tend to form best in moderately to highly resistant rocks where lines of weakness (e.g. joints, cracks, or layers in the rock) are widely spaced, such as in some types of granite, basalt, quartzite, sandstone, and shale. This might seem counterintuitive but it helps to explain why potholes can persist in fast-flowing, erosive rivers. Moderate to high bedrock resistance may make it hard for potholes to form initially and mean that development rates remain relatively slow, but the resistance helps any developing depression to maintain a coherent, roughly cylindrical form. If rock resistance is too low, or too many lines of weaknesses are present, the rims and sides of developing potholes tend to be eroded by the removal of rock fragments or larger blocks of rock and the distinctive roughly cylindrical aspects are lost (Figure 1.13) or the potholes are removed entirely (Figure 1.17). The moderate to high resistance of rocks in which potholes commonly form and persist means that pothole growth rates are likely to be slow, even where there is an abundant supply of erosive sediment particles. Few data exist but pothole growth rates are likely to be significantly less than one millimetre a year. With an average growth rate of 1 mm every year, it would take a pothole 1000 years to deepen and widen by 1 metre. With average growth rates of 0.1 mm and 0.01 mm every year, it would take a pothole 10 000 and 100 000 years, respectively, to deepen and widen by the same amount. In Welsh rivers, modern observations and historical evidence tend to reveal little or no significant change to potholes, even during major floods, which suggest that average pothole growth rates are probably towards the lower end of this range. In most cases, therefore, pothole development is essentially undetectable on a human timescale.

Given the slow development rates, the precise age of many potholes in Welsh rivers is hard to establish with any certainty. But the size of many potholes – tens of centimetres to a metre or more – suggests a very great age for their initial formation, at least thousands of years and quite possibly many tens or even hundreds of thousands of years. Some older potholes thus may have

Figure 1.17: Isolated pothole (centre) on highly jointed volcanic rocks along the Afon Clywedog near Brithdir, Gwynedd. This pothole is unlikely to persist as the joint-bounded blocks of rock are susceptible to erosion during floods (ST).

survived through some of the numerous ice ages that have occurred across parts of Wales and the northern hemisphere in the recent geological past. It also means that some potholes will be significantly older than any major human structures on Earth, as even the earliest known Egyptian pyramids are no more than 5000 years old.

HOW IS POTHOLE DEVELOPMENT LINKED WITH THE FORMATION OF WATERFALLS, RAPIDS AND GORGES?

As outlined above, deviations from the idealised cylindrical pothole form, and partial erosion or decay of potholes, means that many potholes are associated with a variety of other weird and wonderful bedrock sculpted river features. Indeed, in many Welsh bedrock rivers, these bedrock sculpted forms may significantly outnumber well-defined potholes, and include simple potholes with entry and exit furrows, various adornments such as flutes and spirals, compound potholes (potholes within potholes), and breached potholes that in some cases form natural arches (Figures 1.18 and 1.19).

In addition to their visual appeal, on longer timescales, potholes and related sculpted forms are highly significant for river and valley development. Where they form, potholes can be seen as a key part of the river's 'cutting edge', enabling rivers to erode deeper and deeper into resistant bedrock over time. As shown in Figures 1.14, 1.15 and 1.16, neighbouring potholes can deepen and widen and eventually join together. If enough potholes coalesce, then they can contribute to the carving of a deep, narrow channel (technically known as an 'inner channel', 'slot gorge' or 'slot canyon') through the resistant rock. On the walls of the inner channel, slot canyon or slot gorge, partial remnants of these potholes may survive. For example, at the Devil's Punchbowl on the Afon Mynach at Pontarfynach/Devil's Bridge, a glance at the sides of the Mynach slot gorge will reveal many smooth, curved faces on the rock. These faces represent the traces of former potholes that have contributed to formation of the gorge (Figure 1.20).

In many locations, inner channels, slot gorges or slot canyons have a small waterfall or set of rapids at their upstream end, with river flow dropping abruptly into the slot. More generally, many potholes are found in close association with waterfalls, commonly being found near the base and above the lip of waterfalls. Near the base of a waterfall, pothole coalescence and the erosion of larger

Figure 1.18: Bedrock sculpted forms on the Afon Ystwyth, Ceredigion (DR).

blocks of rock by plucking can contribute to the development of a large feature known as a plunge pool. Large plunge pools that are many metres across may also contain individual potholes, although the turbulent

Figure 1.19: Natural arch on the Afon Irfon at Cammdwr Bleiddiad, Abergwesyn, Powys (DR).

water commonly limits their visibility. Over time, as a plunge pool grows in size by widening and deepening, buttressing support for the near-vertical face forming the waterfall is weakened. This weakening of support can lead to periodic collapse of the waterfall face, thereby contributing to a slow, punctuated, upstream retreat of the waterfall. As the waterfall retreats, a steep-sided gorge is left in its wake (Figure 1.21). Above the lip of a waterfall, pothole growth and coalescence can lead to weakening of the rock in the river bed and may also contribute to the upstream passage of the waterfall and the further elongation of the gorge. In these situations too, traces of former potholes may adorn the walls of the gorge.

Figure 1.20: The Afon Mynach at Devil's Bridge, Ceredigion: left) coalescing potholes at the Devil's Punchbowl (at low water, the breached lower walls of the potholes reveal a natural arch); middle) the lower part of the slot gorge immediately downstream of the Punchbowl; right) full view of the slot gorge, showing the curved traces of former potholes and the lowermost two of the three bridges that span the gorge (ST).

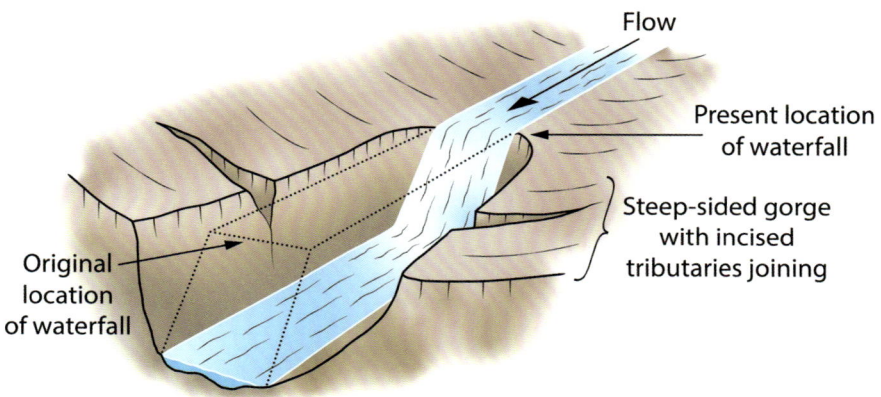

Figure 1.21: Schematic diagram (above) and photograph (below) illustrating how upstream waterfall retreat leads to gorge development (Source of diagram: modified after Hayakawa, Y.S. and Matsukura, Y. (2003). Recession rates of waterfalls in Boso Peninsula. Earth Surface Processes and Landforms, 28: 675-684 and Tooth, S. (2015). The Augrabies Falls region: a fluvial landscape divided in flow but magnificent in spectacle. In Grab, S. and Knight, J. (Eds), Landscapes and Landforms of South Africa. World Geomorphological Landscapes. Berlin-Heidelberg: Springer-Verlag, pp.65-73). The photograph is of the Afon Elan at Pont Hyllfan, Powys (HG).

SECTION 2
• THE ECOLOGY OF POTHOLES •

Section 1 highlighted how potholes are a common feature of many bedrock and mixed bedrock-alluvial rivers in Wales, and outlined how pothole development is related to many other aspects of river scenery, including other bedrock sculpted forms, waterfalls, rapids and gorges. Although pothole development occurs mainly during highly turbulent, sediment-laden flood flows, potholes are more noticeable during more tranquil, low flows, when many are exposed above water level or can be seen through the clearer water. Observations of exposed potholes show that pothole characteristics vary widely, including in terms of size and shape, the amount of infilling sediment, the presence or absence of organic matter such as leaves and twigs, and the levels of water. Some potholes develop biofilms (e.g. composed of algae), unless scoured by the water and sediment (Figure 2.1).

Deeper potholes tend to trap more sediment and organic matter, and may also retain water, even during longer dry spells (Figure 2.2).

Figure 2.2: Organic matter collecting in potholes on the Afon Dulas (north), near Ceinws, Powys/Gwynedd border (ST).

Figure 2.1: Pothole showing biofilm surrounding an area of erosion (DR).

Shallow potholes tend to retain little or no sediment and organic matter, and any remaining water may rapidly evaporate, even if replenished now and then by rainfall.

These diverse characteristics of potholes have important implications for river ecosystems. Where present, potholes increase the diversity of river habitats, benefitting a wide variety of wildlife, including fish, amphibians and invertebrates. Some organisms will use potholes only for brief periods of time, perhaps regularly passing into and out of multiple potholes, whereas others may spend most of their lives inside a particular pothole. This section provides an overview of the ecology of potholes. We hope to demonstrate that potholes are features where it is possible to observe some of nature's most amazing wildlife spectacles, but also that if time is focused on a few select potholes, it is possible to get down close to some equally amazing but underappreciated organisms that may spend a large part of their lives away from the main channel. Peering into potholes is like looking into miniature freshwater worlds, and it can be a very rewarding and enriching experience observing just how much life there is in, and around, these wonderful formations. Nevertheless, as we shall outline, many aspects of river and pothole ecology are sensitive to a variety of human impacts.

FISH

Some of the most dramatic uses of potholes are by migrating salmon. Having spent a large part of their lives out in the North Atlantic, most travel hundreds of kilometres back towards their natal river to spawn. They navigate using only their olfactory system, essentially smelling their way back to their birth places. As they navigate their way upstream, at certain locations on some Welsh rivers, these amazing organisms can be seen leaping out of the water in order to overcome natural barriers such as waterfalls or rapids. While negotiating waterfalls, they often use potholes as places to rest; the plunge pools below waterfalls are also crucial as they need the deeper water to leap successfully. Two excellent places to see leaping salmon are along the Afon Marteg, a tributary of the Afon Gwy/Wye near Rhayader, and at Dolanog on the Afon Efyrnwy/Vyrnwy near Welshpool (both locations are in Powys). Along the Afon Marteg, salmon can be seen attempting to make their way up a cascade of waterfalls in the autumn. The frequent failed attempts are testimony to the arduous nature of this activity. Now and again, a fish can be seen resting in the potholes or plunge pools, halfway up the cascade, before making another attempt to finally overcome the rest of this natural barrier. The strength and agility of these fish is awe inspiring. At Dolanog (Figure 2.3), salmon leap up the lower part of a waterfall and often land in a pothole where they do one of three things: immediately carry on leaping upwards using the momentum of the jump, rest for a bit before leaping again, or let the water carry them back down due to exhaustion. Some salmon make many attempts before succeeding in their quest to scale the full height of the waterfall. Sadly, at Dolanog the salmon cannot overcome the artificial obstacle at the lip of the upper waterfall: a substantial weir. Despite their best efforts, all of them ultimately have to drop back downstream of the waterfall to spawn.

Other species such as trout and eels also make use of potholes. While snorkelling in the Afon Dyfi, we have observed sea trout at the bottom of large potholes pressing their camouflaged bodies against the natural curve of the wall (Figure 2.4) and remaining completely still. During low flows, water currents are negligible at the

Figure 2.3: Salmon leaping from plunge pool to pothole on the Afon Efyrnwy/Vyrnwy at Dolanog, Powys (DR).

bottom of these potholes and it is difficult for predators such as otters to detect them in these locations.

Shoals of fish such as minnows also can frequently be seen in potholes; here, they may be taking advantage of calmer water away from the main flow where they don't need to expend as much energy and possibly also avoiding detection by predators.

Figure 2.4: Sea trout in a pothole on the Afon Dulas (north), near Ceinws, Powys/Gwynedd border (DR).

AMPHIBIANS

Slow moving or static water in potholes that are on the bedrock channel margins, or that are on higher bedrock surfaces and exposed at times of low flows, may be significantly warmer than water in the deeper, faster moving main channel. These warmer water temperatures may have advantages for some forms of wildlife. Although normally associated with pools, ponds and lakes, amphibians can often be seen in river potholes and calmer river reaches in general. In early spring, common toads may use potholes away from the main channel in order to spawn, the double strings of spawn being like a necklace of life (Figure 2.5). A pothole on a bedrock surface isolated from the main flow would also be out of the reach of predators such as fish that would otherwise feast on the spawn and the tadpoles. This does not mean that the spawn and tadpoles are safe from other predators such as newts, which also have been observed in potholes (Figure 2.6).

Figure 2.5: Double string of spawn in a pothole at Pont Llogel near Welshpool on the Afon Efyrnwy/Vyrnwy, Powys (DR).

INVERTEBRATES

By far the most numerous and easily observed organisms

Figure 2.6: Newt and midge larvae cases in a water-filled pothole (DR).

in potholes are invertebrates. From the perspective of invertebrates, each pothole can be viewed as a microhabitat, with different groups utilising different parts of the potholes for varying periods of time. As potholes are natural traps for sediment as well as organic debris such as leaves, twigs and sometimes small branches (Figure 2.7), they are a haven for many invertebrates who rely on these materials for their homes or food.

Figure 2.7: Rich organic debris on a pothole floor (DR).

Some species also feed on the surface biofilms that are composed of microscopic life (e.g. algae) and cling to pothole walls below and above water level, and others live underwater for at least for part of their lives. Other species live on dry land but make use of the water in the pothole every now and then. For instance, many invertebrate species will spend their nymphal or larval stages underwater before embarking on their relatively brief flying adult stages. The most common of these species are collectively known as riverflies and include the mayflies, stoneflies and caddisflies; there are many variations within these broad groups and they can be very abundant in potholes. These species are very important for river food webs, as they are a major source of nutrition for birds such as the dipper, fish such as salmon or young trout, and other insects and arachnids. A noticeable feature of some potholes are the spider webs that are spun across them, a gauntlet of traps that small insect fliers have to negotiate.

Many potholes host a wealth of life not only underwater but also on the water surface; from spring through to mid autumn, many organisms can be seen walking or zooming across the surface. These include the river skaters that are related to the daintier and smaller pond skaters which can also be found in potholes (Figure 2.8); both species have thin, splayed legs that allow them to walk on water. Another surface walker is the water measurer (Figure 2.9), which has a very long head for the size of its body; these are often found on the walls of potholes. Both types of organisms have piercing mouthparts and will attack other, smaller invertebrates and consume their juicy insides using their specialised mouthparts. Skaters will congregate in numbers above potholes at the sides of the main channel and occasionally dart into faster flow. They are very strong and can even be seen during rainfall, seemingly immune to heavy water drops. Each square millimetre of their bodies is covered in thousands of hydrophobic hairs that help them stay dry. River crickets can also walk on the surface and have the ability to secrete a substance that helps break the surface tension of the water so they can move even quicker.

Mayflies is a broad term comprising an order of insects that contains many species. Mayflies, as a general rule,

Figure 2.8: Pond skaters in a pothole on the Afon Claerwen, near Rhaeadr Gwy/Rhayader, Powys (DR).

Figure 2.9: Close-up of a water-measurer (DR).

have three tails which, as larvae, help them to swim. Baetis nymphs (or 'olives') are very good swimmers and can be seen undulating their bodies from one gravel particle to another; they will swim off swiftly if disturbed. They congregate in great numbers and often face into the local current (Figure 2.10). Indeed, their positions can be used to get a general impression of small current movements.

Another common type of mayfly are the Heptagenidae (or stone clingers); these have flattened bodies and are often found underneath the gravel in the bottom of potholes. They are grazers, feeding on algae on gravels and rock surfaces, and their grazing trails commonly provide evidence of where feeding has taken place.

Figure 2.10: Baetis nymphs in a pothole (DR).

Stoneflies are another order of insects commonly found in potholes. They have two tails and crawl among the gravels at the bottom of potholes and on the river bed more generally.

Caddisflies are perhaps some of the most incredible insects that can be seen in potholes. In their larval stage, these underwater architects use natural materials such as sand grains, small gravels, or plant fragments – often in combination – to make cases in which they are protected and camouflaged (Figure 2.11). These cases are sometimes also used as ballast to hold them down in strong currents. They bind the materials together using silk that is secreted from an area near their mouths, and some of their constructions are very elaborate (Figure 2.12). Using natural materials has clear advantages; they blend in very well to their natural surroundings, sometimes their presence only being given away by movement. Empty caddisfly cases are a common sight in many Welsh rivers, including in potholes, but are eventually broken down to once again form part of the natural sediment load. Caddisflies therefore are some of the original recyclers. There are also caseless and retreat- or shelter-making caddisfly species and these tend to be found in faster-flowing river reaches. Among the most remarkable of these species are the net-spinning Hydropsyche (water spirit) larvae that build capture nets out of silk. These can be seen on algae colonising the walls of potholes. Many caddisflies are shredders and help with the breakdown of leaves and twigs that are commonly found in potholes, thereby releasing nutrients for the benefit of the wider river ecosystem.

HIGHER-ORDER VEGETATION AND OTHER LIFE FORMS

In some potholes, particularly those on higher bedrock surfaces isolated from the main flow, a variety of higher-order plants are often present. Moss provides a miniature

Figure 2.11: Caddisflies in a pothole (DR).

Figure 2.12: Empty caddisfly cases (DR).

forest habitat for a wealth of life such as bryozoa (moss animals). In the spring, marsh marigolds are often seen adorning the sides of potholes. In some larger potholes, small willow trees may even be present, having had time to take root between potentially erosive flood flows.

A report from a 1916 field outing of the Builth Wells Naturalists in *The Brecon Radnor Express, Carmarthen and Swansea Valley Gazette, and Brynmawr District Advertiser* highlights the unusual plant life that may grow on boulders around potholes:

> *'Near Penmaenau a band* [of hard rock] *was shot across the river Wye to the Park Wells. The grandeur of the river scenery at this point is due to the difficulty the Wye had in fretting for itself a new course. Some grand pot-holes may here be observed. Hell Hole itself is nothing but a deep pot-hole. The pungent chives, a plant which some observers say was brought by the Romans, grows on the ledges of the river boulders around the pot-holes. The Welsh name for it is 'Syfi Glan Guy* [sic]*' and some writers claim it as the Welsh emblematic flower.'*

HUMAN INFLUENCES ON THE ECOLOGY OF POTHOLES

While many Welsh bedrock and mixed bedrock-alluvial rivers may appear to be in a near-natural, undisturbed condition, there are many human influences on potholes and their ecology. One of the most visible human impacts is the physical pollution that results from materials such as metals, building rubble, and plastics. Some of this material may be deliberately dumped in rivers, or just accumulate in quieter areas away from the main flow, including in potholes.

Other human impacts are less visible. The legacy of mining for minerals such as lead, zinc and silver is still very much in evidence in many parts of the Welsh uplands, with dissolved metals being flushed out of abandoned mines after heavy rain and many sediment particles and bedrock surfaces being stained reddish brown by iron oxides. Agricultural runoff and raw sewage discharges can be another major problem, in some cases leading to discoloured river water and other forms of chemical contamination. These impacts have a negative effect on the ecology of potholes and the wider river ecosystem. Examples include the ingestion of microplastics by invertebrates which then enter the food chain, and the excess phosphorous that can lead to algal blooms and deoxygenation of slow moving or stagnant water.

Many Welsh rivers have also been affected by weirs, dams, reservoirs, and related hydropower schemes. Besides providing the aforementioned impassable obstructions to migrating fish, weirs and dams also disrupt the natural transport of sediment, depriving downstream reaches of important habitat-forming materials. Flow regimes may also be far from natural; the water level in the Afon Rheidol near Aberystwyth, Ceredigion, for example, can fluctuate dramatically over a 24 hour period as water is released from dams for hydropower generation. As levels can rise and fall much quicker than they would naturally, this can impact river ecology, especially in potholes on the channel margins. In other locations, features such as potholes may be hidden for a significant part of the year owing to maintenance of artificially high water levels in reservoirs. The Afon Elan at Pont Hyllfan, near the Elan Valley Visitor Centre provides a particular case in point (Figures 2.13 and 2.14).

Many Welsh rivers that contain potholes, waterfalls and gorges are popular with visitors, who come to enjoy the scenery, experience the ecology, or take advantage of more sporting recreational activities such as kayaking or canoeing. In Section 7, we make our own recommendations for places to visit, but our desire to raise awareness of the majesty of potholes is tempered by realisation that too many visitors can have detrimental effects on the natural environment. A good example is the popular waterfall (and potholes) of Pistyll Rhaeadr, Powys, where many complaints have been made by locals regarding the sheer number of visitors (the narrow approach road has been gridlocked during busy periods) and the amount of litter left behind. Natural river attractions at other tourist hotspots risk being impacted by approach roads and bridges, footpaths, visitor buildings and viewing platforms. Even seemingly innocuous activities such as 'stone stacking' (as a personal marker or form of land art) may be harmful to river ecology because the movement of the gravels may significantly disrupt the habitats for many river species such as invertebrates.

Clearly, it is a fine balancing act between making these environments accessible to the public, with all the benefits for physical and mental health, and protecting the same environments from aesthetic and ecological impacts, particularly the more fragile fauna and flora. Physical and mental health are themes to which we shall return in Section 6.

Figure 2.13: Pont Hyllfan on the Afon Elan, near Elan Valley Visitor Centre, Powys during natural, low-moderate flow conditions (DR).

Figure 2.14: Pont Hyllfan during artificially high flow conditions (DR).

SECTION 3
• POTHOLES AND HISTORY •

Section 2 concluded by discussing the negative impacts that human activities can have on potholes and their fascinating and fragile ecologies. In Wales and farther afield, humans have, of course, lived and worked on the banks of rivers for millennia, and used rivers for provision of water and food, navigation, energy for powering mills, irrigation and other agricultural activities, as well as for various kinds of leisure and recreation. Past human activities have not always had such a significant and lasting impact as many of those activities occurring in contemporary society, but nevertheless may have left their mark in material and immaterial ways. This close relationship between humans and rivers has, over time, provided the setting for historical events as well as inspiration for enduring folk stories, myths and legends. Some of the bedrock and mixed bedrock-alluvial rivers of Wales, with their potholes, plunge pools and deep river reaches, provide excellent examples of where historical events, stories, myths and legends interweave.

FISHING

Anyone with the briefest experience of river angling will know that fish are commonly found in deep, cool pools, often shadowed by trees and away from the reach of predators. As we discussed in Section 2, potholes can also be very important for migrating salmon and trout as they negotiate waterfalls. Historically, this knowledge was exploited by those living on the river banks of rural Wales, particularly working class people, as they tried to supplement their meagre food supplies. Often, this involved making use of the distinct morphologies of potholes. The Welsh artist and travel writer, Edward Pugh, stated in his volume entitled *Cambria Depicta* (1816) that he had visited the waterfall at Dolanog on the Afon Efyrnwy/Vyrnwy (Figure 3.1) and that:

> *'Salmons are caught here, by harpoon-irons being darted at them, on their leaping the rocks.'*

Decades later, other inventive ways of trapping salmon were being employed. For example, Elizabeth Jones, who lived directly above a waterfall at Dolanog, was convicted of salmon poaching. According to a report in *The North Devon Journal* (March 1862):

> *'She pleaded ignorance of the law and the nets were ordered to be destroyed. It appeared the old woman had a peculiar way of taking the fish. Near to her cottage is a shallow waterfall, which the fish, on arriving at the spot, attempted to leap up in order to proceed to their spawning ground. The fish so jumping fell into a hollow in the fall* [pothole]*, where the old woman had previously lodged her net, and on falling into it, they were at once hoisted ashore.'*

Elizabeth's activities were still being recalled decades later in *The Wellington Journal and Shrewsbury News,* which in May 1887 mentions *'Betty Jones's pot'*: evidently, this is referring to the pothole that the salmon leapt into and were trapped. As noted in Section 2, salmon are still seen at Dolanog, albeit in a rather different way, as they are now impeded in their migration by the weir built in the 1920s.

There are other locations along Welsh rivers where

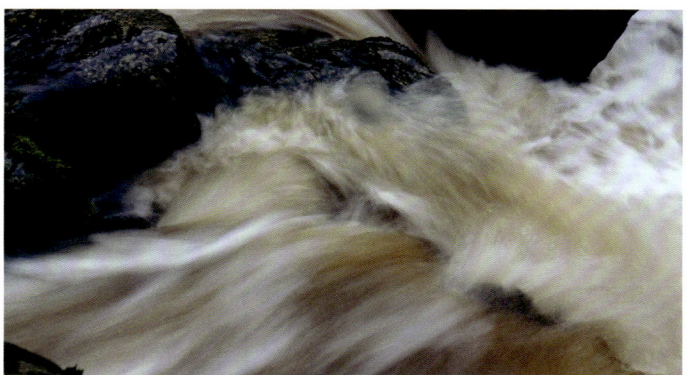
Figure 3.1: Pothole at Dolanog on the Afon Efyrnwy/Vyrnwy, Powys (DR).

people once caught fish by exploiting bedrock geomorphological features, with good examples including the Afon Dyfi near Mallwyd, Gwynedd, and the various rivers in the Afon Conwy catchment. In his book *Dal Pysgod* ('Catching Fish'; 1989), Emrys Evans includes details of many historical practices that made use of river geomorphological features to make fishing more efficient. For example, as in Elizabeth Jones's case, potholes, other sculpted forms and gorges were taken advantage of:

> *'Nets were placed in some instances in gaps created by nature, and in the gorges hewn in the rocks by floodwaters to catch salmon and trout ... Care was taken to weave the nets coarsely enough so that they would not plunder the waters in such a way that meant that were none left to spawn, so that there would be plenty in future.'*

Evans includes the recollections of author Elis Pierce (also known as Elis o'r Nant) regarding the construction and placement of a particular kind of wooden chest (*'cist'*), around two to three yards in length and three to four feet in width, which was driven into the river bed, and designed in a way to trap salmon. River geomorphological features were often used:

> *'In some instances, when convenient, the chest was placed in a neck of a branch of the river shooting out of the main riverbed ...'*

Wooden cages (*'cewyll'*) were also used, and constructed so that they could be placed into bedrock sculpted forms:

> *'The frame was constructed ... to be placed in the opening of the same shape and of the same size as the bed hewn in the rock by the flow. No chisel or hammer or any other tool was used to make any change in the riverbed. The cage was made by hand to fill the bed, whatever its shape or size ... The salmon fell into the cage in a gorge when it tried to leap up the waterfall ...'*

Evans suggests that these methods were long standing in the Afon Conwy catchment:

> *'Grâs Jones, Tan 'R Allt, ... said that her relations had 'lived in Tan 'R Allt' for over three hundred years, but that she had never heard tell of when* [the practice of] *fishing with chests and cages was begun in the gaps, troughs, gorges, beds and sites that were designed and created by the flood waters.'*

Evans also notes examples recounted by the poet, writer and journalist Carneddog, on the Afon Colwyn near Beddgelert, Gwynedd:

'I remember Sara Gruffydd here [living in a house named Cae'r Bompren], a widow living on the parish. She was famous in her day as a fisherman, and one deep whirlpool in the Afon Colwyn is called 'Llyn Sara' to this day. The water would fall over the steep cliff into this pool, and Sara used to place a cage under the cliff. The fish, trying to leap to the top of the cliff, would often fail and slip tidily into Sara's cage.'

While the working classes derived many benefits from potholes and other bedrock river features (Figure 3.2), it seems that the upper classes were not so fortunate. When discussing Rhaeadr Ewynnol/Swallow Falls near Betws-y-Coed, the Ward Lock Red Guide on Conway, Deganwy, Llandudno, North Wales (1921-22) says this of one of the lords of the Gwydyr estate of the Afon Conwy valley:

'There is an old tradition that, as a penance for his oppression of the people, the spirit of Sir John Wynne, of Gwydyr, was doomed to remain in the depths of the pool under the fall, there to be purged and purified.'

Figure 3.2: Potholes at Ffos Anoddun/Fairy Glen, Afon Conwy, Conwy (DR).

DANGERS AND ACCIDENTS

These Welsh bedrock and mixed bedrock-alluvial rivers provided a source of food during hard times, but their steep, smooth and slippery beds and banks and fast currents could be dangerous, even fatal. Searching for mention of potholes in archival material – for example the National Library of Wales's fascinating digitised historical newspaper collection – leads to numerous accounts of tragic accidents, including bodies being recovered from plunge pools and gorges. One such example of a tragic story, told in the *Baner ac Amserau Cymru* newspaper (September 1892) goes into grisly detail:

'The mystery surrounding the disappearance of Mr. William Jones, trader from Tremadog, was cleared up during an inquest held in Lidiart Ysbytty, the home of the deceased, last Thursday.

Shortly after eleven o'clock on Monday, a dark and stormy night, Griffith Williams, farmer, and his two nephews, Richard and Llewelyn, along with a miner, named Stephen, walked from Tremadog towards Cefn Coch Uchaf, along a footpath leading along the edge of the mountain to the uplands. Mr Jones said that the party had a number of items to carry, and took a boy with them, and he accompanied them. When they reached the farm building, some five yards above the waterfall, Mr. Jones bade the party 'Good night' and tried to make his way back through the darkness. That was the last time he was seen alive. Early next morning Griffith Jones, agricultural labourer, saw Mr. Jones's body in a pothole … where the waterfall fell over the precipice. His head had been beaten in between the rocks at the bottom of the pool. There was a three inch wound on his head, down to the bone. Other heavier and lighter wounds were on his face and body. It is thought that the deceased had mistaken the edge of the precipice for the stairs leading down to the footpath, and had fallen to the bottom directly. Death must have been instantaneous. A verdict of 'Accidental death' was returned.'

Even today, serious accidents and occasional deaths occur along the bedrock and mixed bedrock-alluvial rivers of Wales. In August 2021, for example, the Brecon Beacons National Park Authority was urging visitors to take care when visiting the rivers in Waterfall Country (Figures 3.3 and 3.4) after two deaths and a high number of emergency callouts to the area earlier in the year.

FOLK STORIES, MYTHS AND LEGENDS

In many cultures worldwide, folk stories, myths, and legends abound on the banks of rivers and are often associated with the creatures living in them. For example, the cooking of salmon in potholes by the mythical figure of the Coyote features prominently in at least one legend of the indigenous peoples of the Pacific Northwest of the United States; a round-bottomed hole in the rocks along the Big River in Washington State is still called Coyote's Kettle. Similarly, potholes in Wales have been named using words for cooking implements, particularly *'crochan'* (cauldron/pot). For example, William Bingley, an English cleric, naturalist and prolific writer, explored north Wales whilst an undergraduate at Cambridge. In his *A Tour*

Figure 3.3: Potholes and bedrock near Sgwd Isaf, Clun-gwyn on the Afon Mellte, downstream of Ystradfellte, Powys (DR).

Figure 3.4: Sgwd Isaf, Clun-gwyn on the Afon Mellte, downstream of Ystradfellte, Powys (DR).

Round North Wales (1800), he describes potholes he saw in Twll Du (sometimes called the Devil's Kitchen in English) at Cwm Idwal, Eryri/Snowdonia:

> 'Amongst the rocks, at the bottom, I observed a number of circular holes of different sizes, from a few inches in diameter to feet or upwards, which have been formed by the eddy of the torrent from above. These hollows are frequently called by the Welsh people, Devil's pots, and from this circumstance, the place itself is sometimes called the Devil's kitchen.'

Possibly the most famous named pothole in Wales is at Pontarfynach/Devil's Bridge (see Section 1), where one can find the large Crochan y Diafol on the Afon Mynach upstream of the waterfalls (Devil's Cauldron is the direct English translation, although Devil's Punchbowl tends to be more commonly used). Devil's Bridge is an excellent example of where a myth has been inspired by the steep, narrow topography of a bedrock gorge, which has been cut partly through pothole coalescence (Section 1), and possibly also by the toponym Crochan y Diafol, which may have predated the myth. Here it said that a cow belonging to a local lady had somehow crossed the Afon Mynach (Figure 3.5), and while looking for the cow, the devil appeared, and offered to build the lady a bridge across the river to recover the animal. The devil's condition was that he would take possession of the soul of the first living

Figure 3.5: Afon Mynach near the Devil's Punchbowl at Pontarfynach/Devil's Bridge, Ceredigion (ST).

Figure 3.6: View from Pont y Pair, looking upstream, showing potholes and sculpted forms on the Afon Llugwy, near Betws-y-Coed, Conwy (DR).

thing to cross the newly built bridge. In a cunning move, when the splendid new bridge had appeared the next morning, the lady walked down to it with her little dog and saw the devil waiting on the opposite bank. The lady threw a crust of bread over the bridge, which the dog duly chased after. The devil then took possession of the unfortunate dog, rather than the lady, who recovered her cow and walked away having outsmarted the devil.

Another example of these kinds of toponym is the beautiful Pont y Pair (pont = bridge; pair = cauldron) at Betws-y-Coed. The name describes perfectly the broiling water in the Afon Llugwy below (Figure 3.6 and 3.7). Yet another is in the following excerpt from the book by Brian Waters, *Severn Stream* (1949). Blaenhafren means the head or start of the Afon Hafren/Severn and Waters describes in vivid detail his trip to Blaenhafren waterfall (Figure 3.8):

'At present beyond the green mountain slopes there is nothing to distract the walker from the river's beauty, which reaches its artistic culmination at Blaen Hafren. Here the river cascades through a hole in a dome of rock, filling a circular pool, the size of a small room, while another smaller pool lies in the open outside this chamber. Out of the second pool the river flows over a smooth slab of rock twenty-five feet long at an incline of forty-five degrees.

These extraordinary pools formed through the gyrating of stones among the rock are known as Llyn Crochan – the pot pool. I decided to bathe, not without thought that one might be swept out of the pool and down this natural water-chute on to the

Figure 3.7: Afon Llugwy beneath Pont y Pair, looking downstream (DR).

Figure 3.8: Blaenhafren waterfall and plunge pool, Powys (DR).

rocks below. Before entering the water I took a drink at the outer pool to quench my thirst, and then entered by this pool through the narrow portal of the crochan. One is almost deafened by the cascade pouring into these enclosed confines, and has a feeling of imprisonment as the roof half-canopies over one's head, for it is as though one is standing inside an egg, the top of which has been cut open. I stepped over the large loose stone that had created this chamber and plunged into the fascinating obscurity of the place.'

The Afon Conwy catchment is supposedly the home of a mythical monster called the Afanc: variously described as a kind of crocodile, giant beaver or dwarf. It is said to have lived in Pwll (or Llyn) yr Afanc – the Afanc's Pool or Lake – in the Afon Conwy, immediately downstream of the present-day A470 road bridge near Betws-y-Coed and near many striking bedrock reaches (Figure 3.9) such as Ffos Anoddun. The Afanc's thrashing in the Afon Conwy's waters were blamed for the frequent flooding of the valley downstream, and it was decided that the monster had to be removed. A young maiden enticed the monster from the water and the best blacksmith in Wales bound it in iron and chained it to two strong oxen who pulled it all the way to Llyn Glaslyn on the eastern flanks of yr Wyddfa/Mount Snowdon. Such was the effort required by the oxen that the eye of one ox popped out of its socket, and its tears formed another lake: Pwll Llygad yr Ych (pwll = pool; llygad = eye; ych = ox; or Llyn y Foel) near Moel Siabod. The Afanc also features in folklore from Pembrokeshire where the Bedd yr Afanc burial chamber near Brynberian is said to be the resting place of a water monster that had to be slain because of the havoc it wreaked on the local area.

Other mythical beings associated with Welsh bedrock and mixed-bedrock alluvial rivers are the 'Tylwyth Teg' or 'Fair People'; at Pistyll Rhaeadr, the natural arch (see Section 4) was once referred to as a Fairy Bridge. One story, recounted by a correspondent in the *Tarian y Gweithiwr* newspaper (December 1892), is an excellent example of landscape geomorphology figuring in a legend about the Tylwyth Teg, in this case in the upper Afon Conwy catchment (Figure 3.10):

'In Ysbyty Ifan parish, there is a farm called Trwyn Swch; and here, eighty years ago, a young husband and wife lived. They had twins, and one day, with the husband away from home, the wife went out to milk, leaving the twins in the cradle. Who came in but one of the Tylwyth Teg, and in line with their usual tricks, took the babies away, leaving two weaklings of their own offspring in their place. The mother returned, and in her rage after finding the exchange, snatched the aliens in her hands, and rushed with them to the bridge which crosses the terrible gorge of the Conwy, close to the house, and flung them into the whirlpool below. There was a host of Tylwyth Teg there, some of them trying to save their kin, and the others rushing towards the woman. "Grab the witch! Grab the witch!" shouted one of their leaders. "Too late, you clumsy creature!" shouted the wife victoriously from the top of the bank. Some of them ran after her to the house, and in their haste they left behind clear evidence of their existence, and of their habits.

Now, unbelieving correspondents, what do you say, I wonder, when I say to you that this evidence, visible, palpable evidence – is available today? I was almost able to steal one of them to send to the Editor, so that his conviction was sure, but I failed.

Not only do the Tylwyth Teg exist, but they smoke – we saw the pipes that they use! Three or four of them lost their pipes chasing the lady of Trwyn Swch, and they were found there later! The pipes are created beautifully from the blue stone of the gorge, around two or three inches long. Many of them have been found at times close to Trwyn Swch cave, and because they are named 'the pipes of the Tylwyth Teg' and they are related to this folk story, what doubt can there be now about the existence of the Tylwyth Teg.'

Figure 3.9: Pothole on the Afon Conwy near Trwyn Swch, Conwy (DR).

Figure 3.10: Trwyn Swch farm with Dyffryn Conwy in the background (DR).

SECTION 4

• POTHOLES AND THE CREATIVE ARTS •

In Section 3, we discussed how Welsh people have used potholes, plunge pools and other features of bedrock and mixed bedrock-alluvial rivers for fishing. We also outlined examples that illustrate how these features have been associated with danger, and how they have inspired folk stories, myths and legends. In this section, we complement and expand on some of these historical themes by providing select examples of how the visual, auditory and tactile qualities of potholes and related features have inspired various kinds of artistic activity over the past few centuries, including by travel writers and essayists, poets, visual artists and film makers. We begin, however, by considering how potholes can be viewed as natural sculptures, as works of art in the landscape, and why this can provide creative inspiration.

POTHOLE AESTHETICS

As we have shown in preceding sections, the geomorphology, ecology, and social histories of potholes and related bedrock sculpted forms are all capable of inspiring our intellectual curiosity. At a more fundamental level, however, what often draws us to them initially is an appreciation of their aesthetic quality. The photographs in this book illustrate that potholes come in many shapes and sizes and a myriad of types, often along a continuum, commonly is found in the same location. All active potholes – that is, those that are at least occasionally submerged by river flow – are constantly changing, albeit in most cases very, very slowly (Section 1). Flow vortices and the abrasive action of silt, sand and gravel (Figure 4.1) round off jagged corners and smooth rough edges, and over many thousands or tens of thousands of years create the

Figure 4.1: Swirling cobble in a pothole, Afon Dulas (north) on the Powys/Gwynedd border (DR).

Figure 4.2: Underwater sculpted forms on the Afon Twymyn, Powys, with sand and gravel sediment (DR).

Figure 4.3: Underwater sculpted forms on the Afon Twymyn, Powys, with some large cobbles in a pothole (DR).

aesthetically pleasing sweeps and curves of potholes and related sculpted bedrock forms (Figures 4.2 and 4.3).

Appreciation of this 'deeper' (i.e. long) sense of time can help us to see potholes as natural sculptures: works of art in progress. Flow vortices rapidly come and go, and sediment moves spasmodically, so with the passing of seconds, days, weeks, months and years, small changes occur in the potholes, as revealed by percussion marks and chipped fragments. Although most changes are imperceptible on a human timescale (Section 1), with imagination, the cumulative impact of these small changes over longer timescales perhaps can be visualised. Other senses besides just the visual may be stimulated by potholes and related features: the writings of Brian Waters have referred to the auditory experience created by turbulent water running over bedrock (see Section 3) while moving sediment can also create unique sonics. Potholes and other sculpted forms such as furrows, flutes and spirals are also very tactile. They are often very smooth, as are the water-worn pebbles that are found within. Although visiting art galleries can be a hugely rewarding experience, traditionally galleries have been spaces where the tactile experience is discouraged, and where the artwork is roped off or hung on a wall. There are no such restrictions in nature. These visual, auditory and tactile qualities undoubtedly have contributed to potholes and related bedrock features being the inspiration for myth and legend, and for travel and essay writing, poetry, visual artworks, and film making.

TRAVEL AND ESSAY WRITING

The waterfalls of Wales were major attractions for the writers of the eighteenth and nineteenth centuries, many of whom were searching for landscapes that could be regarded as 'picturesque' or even 'sublime', the latter term being used to mean aspects of nature, art, buildings, language, style, or other cultural artefacts that were raised aloft, even exalted. Many waterfalls across Wales came to form part of these writers' 'tours' in which they would often direct the reader to specific places, and sometimes even to specific views, in order that the reader could experience their own sense of the picturesque and sublime. For example, in his *Tours in Wales* (volume 3, 1778), Thomas Pennant, the renowned Welsh naturalist, writer, traveller and antiquarian, describes Pistyll Rhaeadr in Powys (Figure 4.4):

> 'After sliding for some time along a small declivity, it darts down at once two-thirds of the precipice, and, falling on a ledge, has, in the process of time, worn itself a passage through the rock, and makes a second cataract beneath a noble arch which it has formed; on the slippery summit of which, a daring shepherd will sometimes terrify you with standing.'

This natural arch may have been formed by pothole coalescence (see Section 1). Writing many decades later, George Borrow, the renowned travel writer from Norfolk who toured extensively through Wales, took exception to the natural arch. In his *Wild Wales: Its People, Language and Scenery* (1862), he described the arch as an 'ugly black bridge or semi circle of rock' and 'unsightly' and that 'no one could regret if nature in one of her floods were to sweep it away'. Whilst he was standing on 'planks' looking at the waterfall, a woman approached him and introduced herself in 'imperfect English' as the mistress of the house and offered to guide him up the side of the waterfall. He accepted her offer and told her that she could speak Welsh to him. She took him up so close to the falls that he was 'almost blinded by the spray'. The arch now was at their side 'rising like a spectral arch, spray and foam above it, and water rushing below'. He then tells her: 'That is a bridge rather for ysprydoedd [spirits] to pass over than men'. She agrees and adds that she once 'saw a man pass over it' and that he 'wriggled up the side like a llysowen [eel] till he got to the top, when he stood upright for a minute, and then slid down on the other side.'

In his *Guide to the Beauties of Glyn Neath* (1835), William Young relates a journey he took up the Afon Dulais, a tributary of the Afon Nedd/Neath, south Wales, in which he comes across some dramatic potholes:

'The Dylais is one of the most considerable streams which discharge into the Neath, and the ride up the valley through which it flows, will occupy great part of a day: there are many lovely scenes, no considerable falls, but very beautiful rapids; in flowing over the rocky ledges, the action of the water has made some very extraordinary round holes, from one feet to three and four feet in diameter, as though bored by a tool, some of them from five to six feet deep.'

Arthur Aikin, chemist and mineralogist and author of *Journal of a Tour through North Wales and part of Shropshire* (1797), recounts vividly a visit to Pontarfynach /Devil's Bridge. As outlined in previous sections, at this location the Afon Mynach has carved large potholes and a narrow gorge, and then descends over a series of waterfalls to the Afon Rheidol:

'After a long and rather tedious walk ..., we came suddenly to a most singularly striking spot. The valley of the Rhydol contracts into a deep glen, the rocky banks of which are clothed with plantations, and at the bottom runs a rapid torrent. This leads soon to the spot that we were in search of, which is full of horrid sublimity. It is formed by a deep dark chasm or cleft, between two rocks, which just receives light enough

Figure 4.4: Pistyll Rhaeadr, Powys, with its natural arch (DR).

to discover at the bottom, through the tangled thickets, an impetuous torrent, which is soon lost under a lofty bridge. By descending a hundred feet, we had a clearer view of this romantic scene, just above our heads was a double bridge, which has been thrown over the gulph ... The water below has scooped out several deep chasms in the rock, through which it flows before it dives under the bridge.'

Upstream of Devil's Bridge is another bedrock gorge of the Afon Rheidol. This is spanned by a narrow footbridge known as Pompren Ffeiriad/Parson's Bridge and numerous large potholes can be seen (Figure 4.5). Thomas Owen Morgan, writing in his *New Guide to Aberystwyth and its Environs* (1864), described the view from Parson's Bridge 'as a scene both sublime and horrible.'

In their *Gossiping Guide to Wales, North Wales and Aberystwyth* (possibly published around 1921), authors Askew Roberts and Edward Woodall also mention this location but in far more favourable terms:

'The Parson's Bridge is a mile and a half higher up the Rheidol …. It is a magnificent bit of rock and river scenery; and some time might be spent agreeably in exploring the banks of the Rheidol, which rushes through a fine ravine with overhanging rocks, and pot-holes in the bed of the stream.'

POETRY

Many of the descriptions above have a poetic quality, but Welsh bedrock and mixed bedrock-alluvial rivers have also inspired poets themselves. Some of the popular sites

Figure 4.5: Afon Rheidol gorge and potholes at Pompren Ffeiriad/Parson's Bridge, Ceredigion (DR).

previously discussed again figure prominently. Around 1831, Daniel Evans (known also as Daniel Ddu o Geredigion) wrote the following series of englynion, a short four line poem written in the ancient Welsh strict metre of cynghanedd, in which complex rules of rhyme and alliteration create harmony. This poem is provided in full below, along with the English translation, which is based on Richard B. Gillion's (2016) translation. The poem also reflects ideas of the sublime, in that the beautiful yet terrible nature of the awe-inspiring falls and gorge are described, but with respect to potholes and sculpted forms, the description of the flow fashioning the rock into the shape of a *'llwy'* (spoon or ladle) is particularly effective.

I Bont ar Fynach (Yn hon a elwir Pont y Gwr Drwg)

Yn burlan uwch y berwlif - y sefi,
    Uwch safn cwm y mawrlif;
    Lle y llam yr hylla' llif
Ddeugeinllath yn ddig wynllif.

Wyt gadwyn yn dwyn dywenydd, - a chlod
    Gwych lydan trwy'n bröydd;
    A rhaff, na ddaw byth yn rhydd,
Wyd i fwnwgl dau fynydd.

Danat heb baid naid ofnadwy - raiadr
    Gan ruo trwy'r adwy, -
    A chan nerth ei ryferthwy,
Gwna y llif y graig yn llwy.

Mae anian ddiran yn gynddeiriog - wyllt,
    A'i gwedd yma'n llidiog;
    Ei tharan sy'n fytheiriog,
A'i gruddiau yw creigiau crôg.

O'th ben, bont fwynwen fanol, - edrycher
    I drachwith bant ffrydiol;
    Pa enaid na naid yn ol
O sydyn fraw arswydol!

Mi a wn mai rhyw ymenydd - cadarn
    Fu'n rhoi cydiad celfydd
    Y Bont addien ysblennydd,
Uwch ceudod yn syndod sydd.

Llesiol wyt, trwy'r holl oesoedd, - da waith,
    I deithwyr, aml filoedd;
    Camfa lân uwch cwm â'i floedd
Nwyfus yn cyrhaedd nefoedd.

Gwir draws yw rhoi i'r Gwr Drwg - y moliant
    Am haeledd mor amlwg;
    I ni dda mae'n wir na ddwg
Etifedd cas y tewfwg.

To Pont ar Fynach (That which is called the Devil's Bridge)

Purely above the boiling flow - it stands,
    Above the jaws of the valley of the great flow;
  Where leaps the roughest flow
  Two score yards as an angry white flow.

Thou art a chain bearing delight - and brilliant
    Broad praise through our vales;
  As a rope, which will never come free,
  Thou art to the neck of two mountains.

Beneath thee without ceasing a terrible leap - a waterfall
    Roaring through the gap, -
  And with the strength of its torrent
  The flow makes the rock a ladle.

The undivided nature is furious - wild,
    And its countenance here angry;
  Its thunder is threatening,
  And its cheeks are hanging rocks.

From thy head, a gentle, white, detailed bridge - is to be seen
    To a vehement, gushing hollow;
  What soul would not leap back
  From sudden horrific terror!

I know that some mind - firm
    Did put a crafty joint
  The fine, splendid Bridge,
  Above the hollow which is an astonishment.

Beneficial art thou, through all the ages, - a good work,
    For travellers, many thousands;
  A pure step above a valley and its lusty
  Shout reaching heavens.

Truly contrary is giving to the Devil - the praise
    For generosity to evident;
  For our good it is truly not the evil
  Hated legacy of the thick fog.

Devil's Bridge also figures in the work of one of the most famous poets of the Romantic movement: William Wordsworth. Having visited the site, Wordsworth wrote the following poem (1824). 'Pindus' refers to the Pindus mountain range in northern Greece and southern Albania. 'Viamala' refers to a narrow gorge along the Hinterrhein River in Switzerland, which is very similar in appearance to the gorge at Devil's Bridge, featuring potholes and even having been crossed by several bridges historically. In the Romansh language of the region, Via Mala means 'bad path'.

Figure 4.6: Historical photograph (1886) of Pont Hyllfan, Powys, showing the wooden footbridge over the gorge.

To the Torrent at Devil's Bridge
How art thou named? In search of what strange land,
From what huge height, descending? Can such force
Of waters issue from a British source,
Or hath not Pindus fed thee, where the band
Of patriots scoop their freedom out, with hand
Desperate as thine? Or come the incessant shocks
From that young stream that smites the throbbing rocks
Of Viamala? There I seem to stand,
As in life's morn; permitted to behold,
From the dread chasm, woods climbing above woods,
In pomp that fades not; everlasting snows;
And skies that ne'er relinquish their repose:
Such power possess the family of floods
Over the minds of poets, young or old!

On the other side of the Cambrian Mountains, many similarly spectacular bedrock and mixed bedrock-alluvial rivers can be found. William Lisle Bowles was a nature poet and contemporary of Wordsworth, and in his lengthy poem *From Coombe-Ellen* (1798), he describes in vivid terms his visit to Cwm Elan on the Afon Elan in Powys. The following extract most likely relates to a section known as Pont Hyllfan (pont = bridge; Hyllfan = rough rock):

And lo! the footway plank, that leads across
The narrow torrent, foaming through the chasm
Below; the rugged stones are washed and worn
Into a thousand shapes, and hollows scooped
By long attrition of the ceaseless surge,
Smooth, deep, and polished as the marble urn,
In their hard forms. Here let us sit, and watch
The struggling current burst its headlong way,
Hearing the noise it makes, and musing much
On the strange chances of this nether world.
How many ages must have swept to dust
The still succeeding multitudes that "fret
Their little hour" upon this restless scene,
Or ere the sweeping waters could have cut
The solid rock so deep! As now its roar
Comes hollow from below, methinks we hear
The noise of generations as they pass.

Historical photographs show both the 'footway plank' and the gorge in its unmodified state (Figure 4.6), but a stone road bridge has long since replaced the plank (Figure 4.7) and the 'ceaseless surge' has been tamed by the upstream dam. Nevertheless, at times of low water, the 'smooth, deep, and polished' potholes at Pont Hyllfan can still be seen (Sections 1 and 2).

Some Welsh poets ventured farther afield and found inspiration on the banks of bedrock rivers in other countries. Evan Jones (also known as Ieuan Ionawr) published two poems about rivers in South America in *Y Tyst Cymreig* (1869). Extracts of these poems are given below, with the English translation by one of the authors of this book (Hywel Griffiths).

The first poem relates to the Tequendama waterfalls, plunge pools and potholes on the Río Bogota in central Colombia (possibly incorrectly spelt as Sequendama).

Rhaiadr Mawr Sequendama *(detholiad)*
Y ffrwd oedd yng ngwaelodion y crech-lyn yn crych-droi,
Ac amryw o enfysau amryliw'n ei gordoi;
Fel trydan hwy ymlidient y naill ar ol y llall,
A thwf rhuadwy'r weilgi yn berwi oedd ddi-ball;
Y ffrydlif a ymruthrai hyd wely garw'r graig
A chollai'n llwyr ei hunan yn nhrobwll dwfn yr aig:
Ar ochrau'r crwnlyn anferth y crogai llysiau fyrdd
A'r lluoedd heirdd dwmpathau mewn gwyn a choch a gwyrdd;
Y cangau braisg a'r blodau ymblethent bob yn ail
Ac adar paradwysaidd yn dawnsio rhwng y dail,
Uwchben y gwagle erchyll ehedent weithiau'n llon
A'u plyf amryliw euraidd yn annarluniawl bron.
Y ffrydiau byw grisialaidd dan furmur aent i lawr,
Nid oeddynt ond fel defnyn i'r pair berwedig mawr,
Pe rhuai mil o lewod newynog yn eu broch
Eu lleisiau er mor echrys a foddai'r rhaiadr croch;
Ar waelod berw'r ceubwll pelydrai'r haul yn wan,
A chroch daranau'r cwympiad yn crynu'n mhell o'r lan.

The large waterfall of Sequendama (extract)
The flow was in the depths of the bubbling lake, rippling around
And many multicoloured rainbows covering it;
Like electricity they chased each other,
And the noisy torrent boiling grew ever greater;
The flow rushed along the rough, rocky bed
And lost itself in the deep whirlpool of a multitude of waters:
On the edge of the massive, circular lake, a multitude of flowers hung
In pretty clumps, red, white and green,
The stout branches and flowers wove together
With the birds of paradise dancing between the leaves,
Above the terrible void they flew happily
And their golden, multicoloured feathers impossible to draw,
The lively, crystal currents murmured as they fell,
They were nothing but drops in the large, boiling cauldron,
If a thousand hungry lions roared in anger
Their fearful voices would be drowned by the fierce waterfall;
At the bottom of the pothole the sun's rays shone weakly
And the fierce thunder of the falls trembled far from the riverbank.

The second poem relates to a natural bedrock arch on an unnamed river in the Andes:

Y bont naturiol (*detholiad*)
Y bwa clegyrog uwchben y llyn tro
Sydd dri o glogwyni, ac un yn faen clo,
A thrwyddo y gwelir y llynclyn islaw ...

...Y fangre sydd aruthr ac hefyd yn dlos;—
Y pontydd ysgythrog a'r llynclyn a'i wg
Sydd ddelw ddigymhar o drigfa'r gwr drwg.

The natural bridge (extract)
The rocky arch above the whirlpool
Is three cliffs, and one keystone,
And through it we see the whirlpool below ...

The place is terrible, but also fair -
The rough bridges and the scowl of the whirlpool
Is an incomparable likeness of the abode of the devil.

We can speculate as to whether or not Jones's description of this site as the 'abode of the devil' was influenced by the naming of Devil's Bridge in Wales, but in common with many poems, his work succeeds in communicating the sense of the sublime experienced when in the presence of potholes, other sculpted bedrock forms, and cascading water.

VISUAL ARTWORKS

Some painters and photographers have also been inspired by Welsh bedrock and mixed bedrock-alluvial rivers. In comparison to the attention given to waterfalls and gorges as subject matter, however, there has been relatively limited specific focus on potholes and related sculpted forms.

Nevertheless, in the book *Wales: The First Place* (1982), with text by Jan Morris and images by Paul Wakefield, there is a long exposure photograph of the Afon Twymyn showing examples of natural sculptural features and a close-up photo of a pothole on the Afon Lloer in Eryri/Snowdonia. In Dyfed Elis-Gruffydd's *Wales: 100 Remarkable Vistas* (2017), which focuses on 100 of Wales's most spectacular landscapes, there are several photographs of potholes and related bedrock river features. Given the visual aesthetic qualities of potholes and other bedrock sculpted forms, however, it is perhaps surprising that we have not been able to find more examples of visual artworks.

FILM MAKING

The internationally popular, bilingual, noir detective series *Y Gwyll/Hinterland* was filmed in Ceredigion, and the deep river valleys in the uplands contribute significantly to creating the dramatic, haunting atmosphere in which many scenes take place. Pontarfynach/Devil's Bridge features prominently in many episodes, including in the first episode of the first series, during which the detective, Mathias, must climb down into a plunge pool underneath a waterfall to retrieve a body. As outlined in Section 3, tragic accidents have happened often enough on the slippery banks of Welsh rivers. The filming for *Y Gwyll/Hinterland* plays upon this sense of danger while also tapping into elements of the awe and terror experienced by many of the selected travel writers, essayists and poets featured above.

Figure 4.7: Modern view of Pont Hyllfan at low flow (DR).

SECTION 5
• INTERNATIONAL EXAMPLES OF POTHOLES •

As previous sections have demonstrated, potholes are common features of many bedrock and mixed bedrock-alluvial rivers, not only in Wales but also farther afield. While Welsh rivers have some world-class examples of potholes and associated bedrock sculpted features, most have not been subject to detailed scientific study and also tend to be underappreciated by visitors, especially in comparison to larger scale, more immediately visual river features such as waterfalls and gorges. A comparison with better studied and more widely known potholes in other parts of the United Kingdom and overseas can be insightful, not only with respect to the physical characteristics of potholes (e.g. size, shape) but also regarding the wider cultural values of potholes, including how potholes are perceived by society. In this section, we highlight a selection of international examples where potholes are an important, or even dominant, component of river scenery.

OTHER PARTS OF THE UNITED KINGDOM

Many upland rivers in other parts of the United Kingdom are characterised by potholes and other bedrock sculpted forms, which commonly form in association with waterfalls and gorges. Good examples can be found in many parts of the Scottish Highlands (Figure 5.1), the Pennines, Peak District and Lake District of England (Figure 5.2), and the uplands of Northern Ireland. Just as in Wales, however, many of these features 'pass under the radar', with many having escaped detailed scientific study and rarely being landscape attractions in their own right.

MAINLAND EUROPE

Some countries in mainland Europe have started to

Figure 5.1: River Coe near Glencoe village, Scottish Highlands (ST).

Figure 5.2: River Eden near Kirkby Stephen, Cumbria, England (DR).

devote more attention to the value of river potholes, both for scientific study but also as visitor attractions. Spain provides a particularly good case study. Studies of rivers in the mountains of western and central Spain have contributed to key scientific insights into the evolution of potholes in granite rocks, and in recent decades, Spain has made particular strides in nature conservation and geoconservation, including with respect to many sites of important geological and geomorphological heritage ('geoheritage'). Spain is now one of the countries with the highest coverage of protected natural areas in the European Union (25-30% of its territory). Bedrock and mixed bedrock-alluvial rivers are key components of many Spanish landscapes, and so receive various levels of protection owing to their incorporation within national parks, biosphere reserves, World Heritage sites, geoparks and other local and regional initiatives. In the Miño River, northwest Spain, thermal springs form an important tourist attraction but numerous potholes have developed within the granitic bedrock. The role of these potholes in promoting geoheritage sustainability has been investigated. In an article published in the academic journal *Geoheritage* (2017), Miguel Ángel Álvarez-Vásquez and Elena De Uña-Álvarez have argued that potholes have an environmental, didactic, socioeconomic and cultural potential (see further reading in Section 9). They outline how actions have been designed to raise awareness of these values, including educational activities and resources, trilingual (Galician, Spanish, English) posters and leaflets, and self-guided trails.

In other mainland European countries, recognition of the scientific and wider cultural value of potholes and related bedrock sculpted features is also growing. In southern France, for example, many potholes can be found along rivers draining the Pyrenees and the Massif Central, while in northeastern Italy, good examples of potholes can be found along rivers draining the Dolomites.

NORTH AND SOUTH AMERICA

Potholes and related sculpted features are a prominent component of many bedrock and mixed bedrock-alluvial rivers in North and South America. Some have been subjected to scientific study, and the wider cultural values have been recognised and even promoted. For instance, many potholes, related sculpted features, waterfalls and gorges had significance for the indigenous peoples of the Americas; in Section 3, for example, we outlined how the cooking of salmon in potholes by the mythical figure of the Coyote is prominent in at least one legend of the indigenous peoples of the Pacific Northwest of the United States. Furthermore, in many traditional belief systems, entities inhabit waterfalls and plunge pools and guide behaviour around them. In *Myths of the Cherokee* (1902), ethnographer James Mooney documented waterfall Thunder Spirits in some of the legends of the Creek and Cherokee of the southeastern Appalachians of the United States; water monsters living in the plunge pools can upset boats and swallow people who do not acknowledge them. For the descendants of these peoples, the cultural significance of these locations may endure; perhaps there are similarities with how tales of the Afanc or the 'Tylwyth Teg' in Welsh rivers with potholes, waterfalls and plunge pools persist in Welsh folk stories, myths and legends (Section 3). In Section 4, we also outlined how some of the spectacular South American river potholes and related bedrock features attracted the attention of the Victorian era Welsh poet, Evan Jones (also known as Ieuan Ionawr).

Today, many potholed river reaches are popular locations for dipping or swimming, especially during the hot summer months, as well as for other tourist activities. Diverse examples in the United States include Barton Creek in Austin, Texas, and the Pedernales River at Pedernales Falls State Park ~60 km west, where potholes form part of riverscapes sculpted in limestone (Figures 5.3 and 5.4). Along various rivers in New Hampshire, Vermont, and Maine, potholes have formed in bedrock such as granite. In the oft-photographed Antelope Canyon in Arizona, potholes are found in association with a spectacular, intricately sculpted, narrow sandstone gorge ('slot canyon'), the latter forming in part through pothole coalescence (Figure 5.5). In Zion National Park in Utah, the early stages of pothole growth, pothole coalescence and gorge formation can be seen along some of the smaller channels that are gradually etching their courses into sandstone (Figure 5.6).

In Washington State, some of the world's largest potholes can be found in the Channeled Scablands, an area covering up to ~5500 km^2 that was scoured repeatedly by 'megafloods' during the last ice age. Potholes up to 30 m wide and 5 m deep formed in high-velocity floods that surged across basalt terrain following failure of ice dams that had ponded immense volumes of water, and are found in association with other flood-sculpted terrain including canyons and waterfalls (Figures 5.7 and 5.8). Given the size of these potholes, many have been attributed to hydraulic plucking of joint-bounded blocks under the action of powerful kolks (upward spinning turbulent vortices) rather than the more gradual abrasional processes responsible for the formation of many smaller potholes (see Section 1). Some of these potholes retain water for lengthy periods in the otherwise arid terrain. They are sometimes referred to as 'vernal pools' or 'pothole ponds', and support unique ecosystems. These potholes and other flood-related features are promoted as part of geoheritage trails in a network of State Parks and National Natural Landmarks; while their vast dimensions mean that many can only be truly appreciated from an aerial flyover, nonetheless they are large enough to be seen clearly in Google Earth imagery (see, for example, features near Dry Falls at 47.591979°, -119.312733°).

In Canada too, locations with potholes are sometimes promoted as visitor attractions; for example, on Vancouver Island, British Columbia, Sooke Potholes Regional Park, and the smaller Sooke Potholes Provincial Park abutting its southern boundary, both provide opportunities for cliff jumping, swimming and hiking. The potholes were likely formed towards the end of the last age when meltwaters and sediment from receding glaciers scoured the sedimentary bedrock.

ASIA

River potholes can also be found in many Asian countries. Scientific papers on the forms, processes and ecology of potholes have been published based on research undertaken in countries such as India, China, Thailand and Turkey, and in some cases, the wider cultural values of potholes have also been recognised and promoted. The Kukadi River in Maharashtra, India, is widely promoted as 'one of the wonders of nature', with sections of basalt bedrock being characterised by numerous potholes,

Figure 5.3: Barton Creek, Austin, Texas, United States (ST).

Figure 5.4: Pedernales River at Pedernales Falls, Texas, United States (ST).

Figure 5.5: Antelope Canyon, Arizona, United States (DR).

Figure 5.6: Zion National Park, Utah, United States (ST).

Figure 5.8: Pothole in the Channeled Scablands near Dry Falls, Washington, United States (ST). Part of the walls of this pothole have weathered and collapsed, but the feature remains many metres wide and deep.

Figure 5.7: Lookout and signboards at Dry Falls, Washington, United States. Water is retained in the plunge pools and some of the potholes that formed beneath the waterfalls when they were active (ST).

other sculpted forms and gorges. Several notable temples stand on the river's banks, including a temple of the deity Malganga. The locals believe these potholes to be a blessing of Malganga. In southwest China, the Shenyu River valley in Sichuan Province also has many spectacular examples of potholes and other sculpted forms but in this location the features are etched into sedimentary bedrock.

Along the Mekong River, spectacular collections of potholes in sandstone bedrock can be found at a site locally named 'Samphunbok' or 'San Pan Bak', located on the Thailand/Laos border. The name means 'three thousand holes' or 'three thousand shallow lakes'. As in the Channeled Scablands of the United States, many are large enough (up to 30 m wide and many metres deep) to be seen clearly in Google Earth imagery (see, for example, features at 15.799790°, 105.403443°). The development of these potholes is being studied as part of investigations into the long-term development of the Mekong River, and there are plans to promote the Samphunbok site as part of the UNESCO Global Geoparks Network (see further reading in Section 9).

AFRICA

In many African countries, potholes and related sculpted features are prominent along many bedrock and mixed bedrock-alluvial river reaches, and commonly form in association with waterfalls, inner channels, and gorges. Many have diverse cultural values both for indigenous peoples and for later settlers, many of whom were from Europe. At many sites, folk stories, myths and legends abound, and some are major visitor attractions. South Africa provides some of the best examples. For instance, many rivers in the popular Kruger National Park (KNP), eastern South Africa, have carved impressive potholes in various igneous, metamorphic and sedimentary rocks (Section 1), with some having coalesced to form inner channels. Just to the west of the KNP, spectacular examples of potholes in quartzite can be found at the junctions of the Treur and Blyde rivers in the Blyde River Canyon Nature Reserve (Figure 5.9). This area forms part of the Bourke's Luck tourist attraction, named after James Bourke, a gold prospector who staked a claim nearby. Despite the name, Bourke did not find an ounce of gold but later prospectors had better luck, some finding rich deposits in the region. In the Nature Reserve, pothole coalescence has been the primary process by which the rivers have carved inner channels and gorges through the resistant quartzite rock. The remnants of former potholes can be seen on the gorge walls (Figure 5.9) and many potholes at current river level continue to be develop today under high-energy flood flows.

Farther west, the multiple bedrock channels (anabranches) of the Vaal River near Parys, Free State, display many spectacular examples of potholes on granitic and quartzite bedrock surfaces (Figure 5.10; see also Section 1). Farther west still, numerous potholes can be found along the various granite and quartzite bedrock channels that collectively comprise reaches of the Orange River, with many having coalesced to form inner channels. Approaching Augrabies Falls – a series of major waterfalls on the Orange River – some very large examples of potholes can be found in the granitic terrain (Figure 5.11). In *To the River's End* (1948), the journalist and author Lawrence G. Green states that the depth of the plunge pool below the main falls is debated but that it has been reported to be at least 40 m deep. Given that the Vaal and Orange rivers have transported many diamonds across the subcontinent, there have long been rumours that thousands of valuable stones remain trapped in this plunge pool. To our knowledge, none have ever been found in this pool, although diamonds have certainly been recovered from the gravel fills in abandoned potholes in older, fluvially-sculpted terrain near to Augrabies.

Downstream of Augrabies, numerous potholes and associated sculpted terrain and inner channels also can be found along many reaches of the lower Orange River; near the remote Ritchie Falls (!Gariep or Oranje Falls), for instance, large concentrations of potholes are found in association with other sculpted features and inner

Figure 5.9: Bourke's Luck potholes, eastern South Africa (ST).

Figure 5.10: Potholes along the Vaal River near Parys, South Africa (ST).

channels in granite (see Section 1). Some of the unusual formations and deep pools in the river may have helped give rise to the legend of the 'Great Snake' that is rumoured to frequent the lower Orange River. In *To the River's End* (1948), Green recounts how the alleged powers of this snake are many: those who express disbelief suffer

Figure 5.11 One of the authors (HG) contemplating a large pothole near the lip of the main waterfall of the Orange River in Augrabies Falls National Park, South Africa (ST).

ill-health or death, while those who respect the snake can reckon on good fortune, especially among the river diamond diggings. A river ecologist perhaps might be tempted to attribute alleged sightings of the 'Great Snake' to the monitor lizard (Nile monitor, *Varanus niloticus*, also known as the *likkewaan*) that frequents the river and its pools and can grow up to ~2.2 m long. Regardless of where the truth lies, just as in Wales (Section 3) and other countries (for example, see 'North and South America' above), here too the various currents of river geomorphology, ecology, social history and culture can be seen to interweave.

AUSTRALIA/NEW ZEALAND

In Australia and New Zealand, good examples of potholes and related sculpted forms can also be found along many bedrock and mixed bedrock-alluvial rivers. Relatively few have received detailed scientific study, but many are ascribed various cultural values by indigenous peoples or later settlers, some with connections to Wales. For instance, in Fiordland National Park on the South Island of New Zealand, the Cleddau River follows a steep, turbulent course towards Milford Sound/Piopiotahi. The European names derive from the early 1800s when the Welsh sealer John Grono named the sound after Milford Haven in Pembrokeshire, southwest Wales, and the river in Fiordland was also named after a Pembrokeshire counterpart. The road to Milford Sound (State Highway 94) is a spectacular drive popular with tourists, and one short detour takes you to a site called The Chasm, where numerous potholes and a narrow gorge have been eroded into the resistant bedrock (Figures 5.12 and 5.13).

In both Australia and New Zealand, just as in Wales and many other countries worldwide, some river reaches with potholes and other bedrock fluvial features have been damaged, lost, or remain under threat from a variety of human impacts (Section 2). In rare cases, however, the spectacular aesthetics of bedrock sculpted river reaches have helped with conservation efforts. In the 1970s, for example, the Franklin River in the southwest of the Australian state of Tasmania was under threat from dam construction for hydropower. A volume called *Wild Rivers* (1983), with text by Bob Brown and photographs by Peter Dombrovskis (Figure 5.14), showcased the spectacular riverscapes and rainforests along the Franklin and other rivers in the region. After a concerted effort by campaigners, the plans for the dam were shelved; Dombrovskis's dramatic photographs that illustrated what would have been lost beneath the rising waters are widely

Figure 5.12: Potholes and sculpted bedrock forms on the Cleddau River, Fiordland, New Zealand (Tris Irvine-Fynn).

Figure 5.13: Potholes at various elevations along the walls of the Cleddau River, The Chasm, Fiordland, New Zealand (Tris Irvine-Fynn).

Great Ravine' [with a pothole in the foreground]; 'Below the Cauldron, Great Ravine'; and 'Eroded Limestone, Verandah Cliffs'. The photograph 'In Marriotts Gorge, Denison River' shows beautiful sculpted forms, very much like ones seen on the Afon Ystwyth and Afon Mawddach, and 'Pebbles and Pothole' shows a close-up of wetted rock and colourful stones. Finally, 'Pothole in the Denison Gorge' looks down a deep polished pothole.

In view of the threats to many potholed river reaches worldwide, but also growing awareness of their diverse values and associated desire to protect and promote such reaches, perhaps lessons can be learned from this Tasmanian example. For those lucky enough to experience potholed reaches, however, the appeal goes beyond just the purely visual. In *Wild Rivers*, Bob Brown described his journey through the landscape:

'Soon the river narrowed again and the cliffs grew higher and more dramatic. We had entered the Great Ravine, the place of the 'glass-walled cliffs' that had so alarmed earlier parties. We and our craft seemed minute in the immensity of this vault of nature ... For a time the grandeur of this monumental place [Serenity Sound] *flooded my mind. I lost awareness of all else – my raft, my friend, my obligations, myself.'*

regarded as having a highly galvanising impact on the campaign and thus a positive effect on this decision. Photographs of potholes and sculpted forms in the *Wild Rivers* volume include: 'Polished quartzite above Irenabyss'; 'Rock and Rapid below Pine Camp'; 'Deliverance Reach,

In Section 6, we expand on this complex theme of 'awareness' by outlining the wider physical and mental health benefits of potholes and related bedrock features.

Figure 5.14: Front cover of Wild Rivers *(1983).*

SECTION 6

• POTHOLES AND HEALTH •

The words 'potholes' and 'health' initially may not appear to have much in common but in this section of the book we argue that they can be closely linked. These links become more obvious when considered in view of the growing recognition that human wellbeing is partly dependent on a healthy natural environment. Human wellbeing includes the obvious physical health aspects, but also mental health aspects which often can be less immediately apparent. As an example, the term 'nature deficit disorder' has arisen amid growing concerns that humans, especially children, are spending less time outdoors, resulting in a wide range of behavioural conditions. In view of the hard lockdowns or travel restrictions that many people in Wales and farther afield experienced during the first waves of the COVID-19 pandemic (2020 through 2021), many people have become more aware of the importance of spending time outdoors, especially in natural environments relatively free from technological distractions.

Improved physical and mental health is just one of the many benefits that natural environments can provide for us (Figure 6.1). 'Ecosystem services' are defined as all the benefits that people obtain from rivers and wetlands, woodlands and forests, coastlines and oceans, and many other natural environments. The 'Millennium Ecosystem Assessment (MEA)' – a four-year (2001-2005) collaboration involving more than 1360 experts worldwide – provided a state-of-the-art appraisal of the condition of the world's ecosystem services, and assessed the consequences of ecosystem change for human wellbeing. The MEA classified ecosystem services into four main types: *provisioning services* such as food, water, timber, and fibre; *regulating services* that minimise potential damage from climate extremes such as floods and droughts, and improve water quality; *supporting services* such as soil formation, photosynthesis, and nutrient cycling; and *cultural services* that provide recreational, aesthetic, spiritual, and educational benefits. The MEA noted that although the human species is buffered to some extent against environmental changes by culture and technology, it remains fundamentally dependent on the flow of ecosystem services.

POTHOLES AND PHYSICAL AND MENTAL HEALTH BENEFITS

The physical and mental health benefits of being outdoors fall under cultural ecosystem services. Visiting potholes and rivers in general (see Section 7) is a way to access these services. With sensible safety precautions and appropriate permissions from any landowners, visits to potholes, waterfalls, rapids and gorges can contribute to an immersive experience that engages all the senses, with many health benefits.

First, there are the physical health benefits. Some river locations with potholes, rapids, waterfalls and gorges can only be reached by challenging yet rewarding walks. Getting to bedrock and mixed bedrock-alluvial river reaches commonly involves a fair amount of walking on uneven ground, and possibly even steep and vigorous scrambling for the more remote examples (Figure 6.2). This exercise will help develop 'waterfall legs', the added musculature in legs (and arms) that propels someone upwards and downwards, as well as enhance the cardio-vascular system. Stamina is improved the more we

Figure 6.1: Swimming at Camddwr Bleiddiaid, Afon Irfon near Abergwesyn, Powys (DR).

Figure 6.2: Potholes on the remote Afon Paradwys (paradwys = paradise), near Claerwen, Powys (DR).

explore. For the more adventurous, wading or even swimming against the current (when safe to do so) makes us stronger and fitter; swimming in general is a very good form of exercise as it does not put too much strain on the body. Spending time outside, in all weathers, helps build our immune systems too. In Wales and many other countries worldwide, with obesity levels rising, an ageing population, and threat from viruses fresh in our minds, now more than ever it is important to keep physically healthy to enhance personal resilience.

Second, there are the mental health benefits. Studies have shown that even looking at images of nature is beneficial; photos and videos of water are particularly beneficial. But there is growing understanding that actually being in nature provides even greater benefits for mental health as all of our senses come into play: sight, sound, touch, smell, and even taste. In bedrock and mixed bedrock-alluvial rivers, the visual interplay of light and shadow, and the sonics of fast-moving or falling, turbulent water - perhaps accompanied by the rumble of transported gravels - provide unique sensory experiences, as many writers and poets have attested (Sections 3, 4 and 5). Smooth, water-worn rocks and pebbles in and around potholes, and other sculpted forms such as furrows, provide pleasant tactile experiences (Figure 6.3). Even the smell and taste of the spray from potholes, rapids and waterfalls can be enormously enriching experiences. Indigenous cultures in places as far apart as North America and Australia have commonly regarded many potholes, rapids, waterfalls and gorges as special, or even sacred, places (see Section 5). In our modern, often secular cultures, we can still benefit from similar spiritual experiences, as related, for instance, to the aesthetic qualities of potholes or to pothole fauna and flora. Seeing repetition or differences in shapes across a water-worn bedrock surface can give us the same sense of enjoyment as viewing a sculpture, particularly as many bedrock sculpted forms (Figure 6.4) may be highly reminiscent of some of Henry Moore's or Barbara Hepworth's celebrated works. The reflection of a leaf floating on the surface of a half-filled pothole, seeing a pothole bottom through crystal clear water, or watching the stems of a marsh marigold swaying with the breeze, may provide similar enjoyment to looking at a painting in a gallery.

POTHOLES AND OTHER CULTURAL ECOSYSTEM SERVICES

Visiting potholes and rivers in general also has many educational benefits, some of which link in complex ways to improved mental health. Section 10 provides some suggested activities based around potholes that could be undertaken in formal educational settings (e.g. schools, universities) but potholes, waterfalls, rapids and gorges also provide many informal opportunities to learn, irrespective of age. Potholes are a gateway into the world of rivers and nature in general, and there is always something new to see or contemplate, even when re-visiting familiar locations, particularly with the changing seasons (Figure 6.5).

In some cases, we may wish to use mindfulness to help us gain a deeper and richer experience when visiting sites with potholes: focusing just on 'being in the moment', free from other distractions, can be enormously helpful for

Figure 6.3: Smooth, sculpted forms near Betws-y-Coed, Conwy (DR).

putting day-to-day events into perspective and providing a renewed sense of energy to address life's challenges. For example, a sense of awe can come into play when we consider the processes and long timescales involved in the formation and development of potholes and other sculpted bedrock forms (Section 1). Greater awareness of timescales in particular can help us to appreciate the history of the Earth and its natural heritage, and provide a sense of perspective for humanity's own cultural achievements (Section 1). Similar awe can be invoked when recognising the role of potholes in river ecology (Section 2). Looking at the salmon leaping up seemingly insurmountable obstacles such as the waterfall on the Afon Marteg, for instance, can provide awareness of the lengthy timescales of biological evolution that has provided these species with power and perseverance. Salmon are fusiform-shaped because natural selection has opted for highly sleek and streamlined bodies that are able to cut through the water more efficiently. Going against the current is much harder but has its rewards. We can take something from this lesson, just as we can when contemplating the intricacies of caddisfly larvae building cases out of sand, small gravels, and plant material (Section 2).

These and many other lessons may also provide an opportunity to contemplate the more unwelcome human impacts on potholes and rivers, whether direct or indirect, deliberate or inadvertent (Section 2). Evidence shows that people tend to act in more environmentally-friendly ways when they feel connected to nature, and only by becoming aware of our negative impacts can we start to take steps to address them. Visits to bedrock and mixed bedrock-alluvial rivers therefore help to develop a mindset where potholes are seen not only as important for our health, but also as features that are an important part of our wider natural and cultural heritage, worthy of our respect and protection.

Figure 6.5: Sets of rapids through the changing seasons near Pont Llogel on the Afon Efyrnwy/Vyrnwy, Powys. Clockwise from top left: summer, autumn, winter, spring (DR).

Figure 6.4: Sculpted forms on the Afon Dulas (north), near Ceinws on the Gwynedd/Powys border (HG).

SECTION 7
• VISITING WELSH POTHOLES •

As Sections 1 through 5 have illustrated, Wales is blessed with a wealth of river potholes and related bedrock sculpted forms that complement and even rival the splendour of those found internationally. There are many locations in Wales where it is possible to access potholes relatively easily and observe and experience them at close quarters, with all the attendant physical and mental health benefits (Section 6). Access can vary with the seasons, and especially with changing flow levels, but there is something to be seen along potholed reaches of rivers throughout the year and in all weathers. From a health and safety perspective, it is easier to visit potholes during lower flow on a sunny day when more potholes will be visible and accessible (Section 1), and when the natural light will illuminate other river bed features, including those related to the ecology of potholes (Section 2). During higher flows, some potholes will be submerged and access will be more restricted, but at these times we can observe and perhaps better understand the power of the river to create and modify bedrock features.

As there are so many locations to choose from in Wales, what follows are selected examples from different regions showing the range of forms that potholes can take. Some are located close to a layby or car park but reaching others involves walks of varying lengths. We include grid references to enable the locations to be found on Ordnance Survey (OS) maps or apps (a table at the end of the section also provides the *what3words* equivalents) but when visiting the locations it is also advisable to carry a detailed map or use those provided on signboards. Where we mention the left or right bank we are referring to the 'true left' or 'true right', that is, the left and right banks as you look downstream.

A NOTE ON HEALTH AND SAFETY

The dangers of steep river banks, slippery rocks and fast-flowing water mean that extreme caution is needed when visiting potholes (for historical accidents related to potholes, see Section 3). Care must be taken on the approach to, and return from, locations. Most of the selected locations are ones in which potholes may be seen from footpaths or bridges, but even well-trodden footpaths and bridges may be uneven and/or slippery, while rockfalls and landslides can sometimes occur on steep valley sides, especially during or following wet weather. Some images in this book are from locations that are only accessible by relatively arduous gorge walking and swimming upstream, and these locations are not included below. If, however, you do venture into the river at any point, bear in mind that: the strength of the current is often deceptive, even at low flow; water can be deeper than it first appears, especially if the water is clear; and water temperatures, even in summer, can be very low. Many underwater bedrock features are likely to be covered in a surface biofilm (Section 2) that can make them extremely slippery. Be aware that mobile phone signal may not be available at many locations, particularly the more remote ones. As with any adventurous activity, it is always advisable to let someone know of your plans and itinerary.

To the best of our knowledge, all information is correct at the time of writing. The authors do not take responsibility for personal accidents, or damage or loss to personal property while visiting these locations. Please follow the Countryside Code when visiting these locations. Respect local residents and landowners, and take only photographs while leaving nothing but footprints. Please refrain from making your own rock piles (or at least dismantle them before leaving) as they can detract from the pleasure of others and the location itself.

ERYRI/SNOWDONIA

Afon Llugwy, Betws-y-Coed

This is one of the best locations to see potholes in Wales, and one of the easiest to access. Examples can be seen from Pont y Pair in the village centre (see Section 3), where adjacent parking can be found for a fee [SH 791 567]. There is an option to follow a riverside walk along the Afon Llugwy upstream to Miner's Bridge [SH 779 569], where more potholes can be seen. You can then loop back or, alternatively, continue to the renowned Rhaeadr Ewynnol/Swallow Falls, where an entrance fee applies [SH 765 577].

Afon Conwy, Afon Lledr, Ffos Anoddun/Fairy Glen

Although Ffos Anoddun is within walking distance of Betws-y-Coed, parking is also available nearby [SH 798 546]. After paying the entrance fee, it is a short walk to Ffos Anoddun [SH 801 542]. You can proceed downstream towards the Afon Conwy's confluence with the Afon Lledr [SH 798 542] where there are wonderful examples of potholes, particularly at low flow (Figure 7.1). A short drive from Ffos Anoddun is Rhaeadr y Graig Lwyd (Rhaeadr = waterfall; [C]raig = rock; [L]lwyd = grey, also known as Conwy Falls), where more potholes can be seen. Parking is available near the A5 [SH 810 535] and there is an entrance fee.

Figure 7.1: Afon Conwy downstream of Ffos Anoddun/Fairy Glen (DR).

Nant Cwm Llan, Nantgwynant

Parking is available near the A498 for a fee [SH 627 506]. After a short and highly scenic walk along the start of the Watkin path to Cwm Llan, you can take a right fork to follow the path over a slate bridge [SH 622 516]. In the beautifully crystal clear (and very cold!) water, there are amazing potholes located upstream of small waterfalls (Figure 7.2). If you are feeling adventurous, you can rejoin the Watkin path and continue up to yr Wyddfa/Mount Snowdon. Alternatively, you could climb Lliwedd and then drop down to Cwm Llan. Another option for ascending yr Wyddfa after visiting the falls and potholes on Nant Cwm Llan is to follow a footpath west towards the ridge [leaving the Watkin path at SH 621 520] and then proceed northwards along the narrow ridge of Allt Maenderyn.

Afon Mawddach, Afon Gamlan, Ganllwyd

Free parking can be found adjacent to the A470 [SH 726 243]. You can take the footpath to the Afon Mawddach and its confluence with the Afon Gamlan and cross a footbridge: many potholes are visible along this reach. You can then follow the forest track heading northeast then north on the left bank of the Afon Mawddach towards Rhaeadr Mawddach and Pistyll Cain; a circular route is made possible by returning on the right bank. Many potholes can be seen from both sides of the river and especially from bridges at low flow (Figure 7.3). There is additional parking on the west side of the river [SH 733 251 and SH 735 263 – the latter is approximately 1 km south of the waterfalls]. Gold was mined in this area and some workings are still visible at Gwynfynydd. Rhaeadr Du on the other side of the A470 is also very much worth a visit; to get there, you can follow the path on the left bank of the Afon Gamlan.

Figure 7.2: Nant Cwm Llan (DR).

Figure 7.3: Afon Mawddach (DR).

CEREDIGION AND MALDWYN (MONTGOMERYSHIRE)

Afon Efyrnwy/Vyrnwy, Pont Llogel

Free parking is available at the Natural Resources Wales car park/picnic site [SJ 032 154] and then you can follow the riverside walk downstream. If the flow is low, many potholes will be visible on the river bed (Figure 7.4). At the second set of small rapids [SJ 042 145], some excellent examples can be seen. You then have the option of continuing to Dolanog where additional examples are located on a set of rapids downstream of the village [SJ 071 129]. In the autumn, migrating salmon and trout can be seen leaping at the falls [SJ 067 126]. A longer but lovely walk can be taken from Pont Llogel, passing Dolanog, and following parts of Glyndŵr Way to Pontrobert; to complete the circular walk, follow the road northeast out of the village and then take the first left. After about 2 km take a left track into a forest. This path follows part of the Ann Griffiths' walk. By following this track and then a lower one, you will reach Dolanog. Another option is to take a vehicle to Dolanog and explore from there.

Afon Banwy, Llanfair Caereinion

During periods of low flow, many potholes can be seen from the B4385 road bridge over the Afon Banwy, which is also known as the Afon Einion (see front cover) [SJ 104 065], and from a riverside walk in the nearby Deri Woods [SJ 099 066].

Afon Twymyn, Dylife

A very short walk from roadside parking [SN 871 939] takes you to a secluded spot on the Afon Twymyn upstream of Ffrwd Fawr waterfall, which is one of the highest in Wales. Here, many bedrock sculpted forms (including potholes) can be seen in water that is crystal clear for much of the year (Figure 7.5). These sculpted forms can be seen a safe distance upstream from the lip of Ffrwd Fawr waterfall, and we strongly advise against getting close to this slippery and hazardous spot. Nearby, to the east, there is a layby adjacent to the road at SN 873 939 where stunning views of the dramatic Twymyn gorge can be seen.

Figure 7.4: Afon Efyrnwy/Vyrnwy (DR).

Figure 7.5: Afon Twymyn (DR).

Afon Rheidol, Cwm Rheidol

You can use the roadside parking at the end of a narrow lane [SN 732 779] that leads you upvalley from Cwm Rheidol hydroelectric power station. Many potholes of varying sizes can be seen in the nearby rapids (Figure 7.6). Remains of the metal mining industry can be seen on the river banks and valley sides. When driving back, you can park opposite Cwm Rheidol power station and walk a short distance upstream to the picturesque Rheidol Falls [SN 709 789]. Some potholes are visible from above the falls at low flow. Salmon can sometimes be seen leaping here in the autumn.

Afon Rheidol, Pompren Ffeiriad/Parson's Bridge

A short walk from the layby at Ysbyty Cynfyn [SN 752 790], located on the road between Ponterwyd and Pontarfynach, takes you through woodland and to a high footbridge over the Afon Rheidol. Very large potholes can be seen both upstream and downstream from the high vantage point of the footbridge [SN 748 790] (see Section 4). The remains of the historic Temple metal mine are nearby.

Afon Mynach, Pontarfynach/Devil's Bridge

This location is a short drive downvalley from Pompren Ffeiriad/Parson's Bridge. Parking at the falls is free [SN 741 770] but there is an entrance fee to the upper section of the falls (Devil's Punchbowl) and to the lower section. If the kiosk is not open, pound coins are needed for the historic coin-operated turnstiles on the upper section. On the upper section, there are two very large, breached potholes and great views of the narrow gorge and the three bridges that span it (see Sections 1, 3 and 4). On the lower section,

Figure 7.6: Afon Rheidol (DR).

you can descend to the bottom of the tiered waterfalls. Paths can be very slippery, and special care is needed on the final descent down Jacob's Ladder, which consists of many steep, narrow steps. The different sections of the waterfalls and associated plunge pools can be viewed on the ascent back to the road and the Hafod Hotel.

Afon Ystwyth at Hafod, near Pont-rhyd-y-groes
Free parking is available at the Natural Resources Wales car park adjacent to Hafod Church [SN 768 736]. There is a map at the car park and from here, it is a short walk to a footbridge over the Afon Ystwyth gorge (only 2 people at a time!). Stunning examples of potholes can be seen from both sides of the river (Figure 7.7).

Afon Ystwyth, Pont-rhyd-y-groes
You can park in the village near the historic working waterwheel [SN 738 722] and follow the nearby path down to a footbridge over the Afon Ystwyth. If you cross the footbridge, you can then follow the path and walk westwards and downstream through Coed Maenarthur woodland. Potholes are visible along the river between the footbridge and the end of the woodland (see Section 1), particularly near a waterfall that is approximately 1.5 km downstream.

Afon Teifi, Henllan
This location is reached by turning off the A484 onto the B4334 and parking on the roadside on the Ceredigion side of the bridge [SN 355 400]. If you follow the footpath upstream on the right bank of the river, excellent examples of potholes are visible between the footpath and the main channel of the Afon Teifi as it flows through Henllan gorge.

Afon Teifi, Cenarth
Here, you can park for a fee at Cenarth on the Ceredigion side of the bridge [SN 269 416]. Potholes are visible between the car park and the main river channel, and you can walk upstream to enjoy stunning views of the rapids.

Figure 7.7: Afon Ystwyth gorge at Hafod (HG).

Figure 7.8: Afon Marteg (ST).

MID AND SOUTH POWYS AND CARMARTHENSHIRE

Afon Marteg, Gilfach Nature Reserve

This location is reached by following the signpost from the A470 to the free car park [SN 953 714]; from here, a riverside walk along the Afon Marteg leads to a waterfall. Potholes can be seen along various bedrock sections and especially at the lower falls when water levels are low (Figure 7.8). There are other walks in this area, including one onto the hillsides on both sides of the Afon Gwy/Wye, one of which takes you into the upper Cwm Elan. Nearly opposite the entrance to the reserve, there is a path from a layby to a footbridge over the Afon Gwy/Wye, from which more potholes may be seen.

Afon Gwy/Wye, Rhaeadr Gwy/Rhayader

You can park at the small free car park (on the road to Cwm Elan) or on the roadside: both parking spots are in Rhayader itself [SN 968 678]. A short distance from the car park, a path leads you down towards the river, and here a number of potholes, including some large examples, can be seen.

Figure 7.9: Afon Gwy/Wye at Rhaeadr Gwy/Rhayader (DR).

Figure 7.10: Afon Claerwen (DR).

Afon Elan, Pont Hyllfan

You can park for free at the small Dŵr Cymru Welsh Water car park near Pont Hyllfan [SN 914 672], located on the road between Garreg-ddu and Penygarreg reservoirs, and then walk to the small bridge nearby. When water levels (controlled by dam releases) are low, this is one of the very best locations to see potholes in the British Isles (see Sections 2 and 4).

Afon Claerwen and Afon Arban near Claerwen dam

You can park at the side of the road leading to Claerwen dam [SN 885 626] to view potholes at and downstream of the waterfall (Figure 7.10). Further up the road, and just downstream of Claerwen dam, are beautiful potholes, including breached and coalesced examples. These are easily seen from the bridge [SN 869 634] when water levels are low. Numerous excellent potholes also can be seen on the nearby tributary, the Afon Arban, including a number from the footbridge.

Afon Irfon, near Llanwrtyd

From the A483 at Llanwrtyd Wells, you can follow the signs for Abergwesyn; a few kilometres up the increasingly narrow road there are laybys either side of the road at a site called Pwllgolchi [SN 859 499]. Here, the Afon Irfon has carved a miniature gorge with some lovely potholes. A short distance along the road, Pwll Bo free car park is on the right [SN 856 507]. More excellent examples of potholes can be seen from the bridge, and you can look out for other potholes as you travel along the road.

Afon Irfon, near Abergwesyn (Camddwr Bleiddiaid)

A few kilometres along the stunning mountain road to Abergwesyn, you can park on the roadside [SN 839 550]. This reach of the river has numerous superb examples of potholes and other bedrock sculpted forms, including natural arches and tunnels. This is Camddwr Bleiddiad (Camddwr = crooked water; Bleiddiaid = wolves) – a stunning, rocky, narrow and, in places, very deep reach (see Sections 1 and 6). Some parts are so narrow, it appears possible (but not advisable!) to leap across.

Upper Afon Tywi

This location can be reached by proceeding along the road from Camddwr Bleiddiaid towards Tregaron; roadside parking allows views of the upper Tywi and, at low flow, more excellent potholes can be seen both upstream and downstream of the bridge [SN 803 571] (Figure 7.11). You can then take the road south to Llyn Brianne and head to Dinas Nature Reserve, which has free parking [SN 787 470]. This is near the confluence of the Afon Pysgotwr and the Afon Tywi. Here, there is a riverside walk with many potholes normally visible.

Figure 7.11 Upper Afon Tywi (DR).

Bro'r Sgydau/Waterfall Country

The biggest concentration of waterfalls (and arguably, of potholes) in Wales is in an area of the Brecon Beacons National Park known as Bro'r Sgydau or Waterfall Country. There are several named waterfalls in this area as well as many smaller, unnamed waterfalls. The two main starting points for walks are near Ystradfellte [SN 929 134] and Pontneddfechan [SN 900 075]. If flow is low, you will see dramatic potholes at several sites. There is a Brecon Beacons National Park car park at Cwm Porth [SN 929 124]; a fee is payable, yet this can still get very busy. A free alternative is a small car park near Clun-gwyn [SN 918 105] but this also can get busy. Overspill car parks do come into operation in the summer. Waterfalls and potholes can be found on the Afon Mellte, Afon Hepste, Afon Pyrddin and Afon Nedd Fechan (Figures 7.12, 7.13 and 7.14). Some of the most dramatic potholes are located at Sgwd Isaf at Clun-gwyn. At nearby Sgwd y Pannwr, mating toads can be seen in some of the large potholes in the spring.

Figure 7.12: Afon Mellte (DR).

Figure 7.13: Afon Hepste (DR).

Figure 7.14: Afon Nedd Fechan (DR).

Location	Grid reference for general location	what3words for general location
ERYRI/SNOWDONIA		
Afon Llugwy, Betws-y-Coed	SH 791 567	fattening.spaces.sits
Afon Conwy, Afon Lledr, Ffos Anoddun/Fairy Glen	SH 798 546	singer.reapply.gratuity
Nant Cwm Llan, Nantgwynant	SH 627 506	bowls.sparrows.watchdogs
Afon Mawddach, Afon Gamlan, Ganllwyd	SH 726 243	renting.thrusters.nimbly
CEREDIGION AND MALDWYN (MONTGOMERYSHIRE)		
Afon Efyrnwy/Vyrnwy, Pont Llogel	SJ 032 154	befitting.response.metro
Afon Banwy, Llanfair Caereinion	SJ 104 065	nosedive.decisions.movie
Afon Twymyn, Dylife	SN 871 939	heartless.refreshed.hazelnuts
Afon Rheidol, Cwm Rheidol	SN 732 779	slept.cowboy.dices
Afon Rheidol, Pompren Ffeiriad/Parson's Bridge	SN 752 790	waged.loitering.then
Afon Mynach, Pontarfynach/Devil's Bridge	SN 741 770	grudging.property.sleeping
Afon Ystwyth, Hafod, Pont-rhyd-y-groes	SN 768 736	hotels.ears.playfully
Afon Ystwyth Pont-rhyd-y-groes	SN 738 722	thumbnail.milkman.heaven
Afon Teifi, Henllan	SN 355 400	firework.awards.melon
Afon Teifi, Cenarth	SN 269 416	rejoins.thud.required

Location	Grid reference for general location	*what3words* for general location
MID AND SOUTH POWYS AND CARMARTHENSHIRE		
Afon Marteg, Gilfach Nature Reserve	SN 953 714	mourner.blubber.resonates
Afon Gwy, Rhaeadr Gwy/Rhayader	SN 968 678	vowel.embraced.jets
Afon Elan, Pont Hyllfan	SN 914 672	slipped.tailors.proclaims
Afon Claerwen and Afon Arban	SN 885 626	broached.woodstove.gives
Afon Irfon, near Llanwrtyd	SN 856 507	titles.trappings.evenings
Afon Irfon, near Abergwesyn (Camddwr Bleiddiaid)	SN 839 550	auctioned.decimal.broken
Upper Afon Tywi	SN 787 470	takeovers.original.upcoming
Bro'r Sgydau/Waterfall Country	SN 900 075	these.even.shortcuts

SECTION 8
• GLOSSARY •

The geomorphological terms commonly used to describe the initiation and development of potholes, and river processes and forms in general, may be unfamiliar to some readers. Below are some brief, simplified definitions of some of the most important terms used either in this book (Sections 1 and 5 especially) or in other river geomorphology texts or online resources (see Section 9). The list isn't exhaustive, but should provide a good grounding for understanding the key processes and landforms associated with different river types in Wales and farther afield.

Abrasion
A process of erosion in which bedrock surfaces are worn away by the action of transported sediment; abrasion generally makes surfaces smoother (Figures 8.1, 8.2, 8.3 and 8.4).

Figure 8.1: Abrasion and attrition in action in a pothole on the Afon Efyrnwy/Vyrnwy at Pont Llogel, Powys (DR).

Alluvium
Loose sediment that has been eroded and transported by rivers; it can contain a variety of materials such as fine particles of clay and silt as well as larger particles of sand and gravel.

Attrition
A process of erosion where pebbles and other sediment particles strike against each other and the river bed, grinding or breaking the particles down into smaller pieces (Figure 8.1).

Bedrock
The relatively hard, solid material at or near the earth's surface; bedrock can be igneous, metamorphic or sedimentary.

Bed load
Particles in a river that are transported along the bed; bed load moves by rolling or sliding (traction) and/or by hopping or bouncing short distances (saltation).

Cavitation
Erosion of fragments of rock through the action of shock waves generated by imploding air bubbles at the base of the flow.

Clasts
Discrete fragments of minerals and rock; used in a similar way to sediment particles but most commonly with regard to gravel particles such as pebbles or cobbles (Figure 8.2).

Corrosion
Chemical action in which water dissolves minerals in bedrock and sediment and carries the material away in solution.

Figure 8.2: Sediment and clasts of various sizes on the bed of the Afon Clarach, Ceredigion (DR).

Deposition
The laying down of sediment initially transported by a river (e.g. in slower moving flow or stagnant water).

Erosion
A set of processes in which earth materials (bedrock, sediment) are removed and transported elsewhere; types of river erosion include hydraulic action, abrasion, attrition and corrosion.

Furrow
A longitudinal erosional feature in a river; may develop from, or into, a pothole.

Hydraulic action
A process of erosion involving the power of flowing water e.g. through rapid changes of pressure in fast-flowing water that help to weaken or dislodge sections of jointed or fractured bedrock.

Igneous
A type of rock formed when molten rock (magma or lava) cools and solidifies. Examples include basalt and granite.

Metamorphic
A type of rock that has been changed from its original form (igneous or sedimentary) by intense heat or pressure. Examples include slate, gneiss and marble.

Plucking
A process of erosion (also known as quarrying), mainly

Figure 8.3: Underwater sculpted forms on the Afon Twymyn, Powys (DR).

Figure 8.4: Layers in sedimentary rock, underwater in the Afon Hafren/Severn, Powys (DR).

involving hydraulic action, in which the river current dislodges and removes sections of bedrock; the process is facilitated by lines of weakness formed by joints or fractures (Figure 8.5).

Plunge pool
A pool at the bottom of many waterfalls formed by hydraulic action (possibly including plucking) of the descending and swirling flow, and abrasion resulting from transported sediment.

Pothole
A roughly cylindrical form eroded into bedrock (or sometimes deposits such as clay) in the river bed or banks. Potholes are formed by a variety of erosional processes, most notably abrasion.

Saltation
A process of river transportation where sediment particles hop or bounce short distances along the river bed.

Sculpted forms
A general term for the wide range of erosional features that form and develop in bedrock rivers; examples include potholes, furrows and plunge pools (Figures 8.3 and 8.4).

Sediment
Earth surface material that is transported from one location to another; it can consist of rocks and minerals, and remains of plants and animals, and can range in size from clay particles to boulders (Figure 8.2). Sediment also includes dissolved material transported in solution.

Sedimentary
A rock type formed from the deposition of sediment particles (rocks, minerals, plant or animal remains) or from chemical precipitation of material in solution. Commonly, sedimentary rocks are formed in layers called beds or strata. Examples include sandstone, mudstone and limestone (Figure 8.4).

Suspension
A process of river transportation in which smaller particles are carried along in the current without coming into contact with the river bed.

Solution
A process of river transportation where rivers carry dissolved chemicals.

Traction
A process of river transportation in which larger sediment particles slide or roll along the river bed.

Transportation
The movement of sediment along the river; it includes the physical processes of traction, saltation and suspension and the chemical process of solution.

Weathering
The breakdown of rock at or near the Earth's surface by the action of weather (e.g. wind, rain, extremes of temperature) and biological activity (e.g. growth of tree roots). Rock breakdown can involve purely physical weathering, purely chemical weathering, or some combination of the two.

Figure 8.5: Evidence of plucking (left) on a bedrock outcrop in the Afon Mawddach near Ganllwyd, Gwynedd (DR).

SECTION 9

• FURTHER READING AND ONLINE RESOURCES •

We hope that this book has provided a glimpse into the fascinating world of potholes and their significance for river landscape development, ecology, culture and health. For those wishing to find out more about the natural and cultural aspects of potholes, the publications listed below may be of interest. Many are specific to potholes and related bedrock sculpted forms and are quite technical, but we also include some more general overviews of rivers and geomorphology. Some of these resources may help support the suggested educational activities that we outline in Section 10.

Potholes and geomorphology

The following volume was one of the first collections of scientific papers that focused specifically on bedrock river processes and forms, including potholes and related bedrock sculpted features:

- Tinkler, K.J. and Wohl, E.E. (Eds) (1998). Rivers Over Rock: Fluvial Processes in Bedrock Channels, Geophysical Monograph Series, Vol. 107. American Geophysical Union: Washington DC, 323 pp.

The following volume provides a comprehensive overview of potholes and other bedrock sculpted forms, and refers to many older scientific studies:

- Richardson, K. and Carling, P.A. (2005). A Typology of Sculpted Forms in Open Bedrock Channels, Special Paper 392. Geological Society of America: Boulder, Colorado, 108 pp.

Examples of more recent (i.e. post-2000) specialist studies on a range of geomorphological and cultural aspects of potholes from various locations worldwide include:

- Álvarez-Vásquez, M.A. and De Uña-Álvarez, E. (2017). Inventory and assessment of fluvial potholes to promote geoheritage sustainability (Miño River, NW Spain). Geoheritage, 9: 549-560.

- Ji, S., Li, L. and Zeng, W. (2018). The relationship between diameter and depth of potholes eroded by running water. Journal of Rock Mechanics and Geotechnical Engineering, 10: 818-831.

- Kale, V.S. and Joshi, V.U. (2004). Evidence of formation of potholes in bedrock on human timescale: Indrayani river, Pune district, Maharashtra. Current Science, 86: 723-726.

- Odhiambo, B.D.O. and Manuga, M. (2017). Tshatshingo Pothole: a sacred Vha-Venda place with cultural barriers to tourism development in South Africa. African Journal of Hospitality, Tourism and Leisure, 6: 12 pp.

- Ortega, J.A., Gómez-Heras, M., Perez-López, R. and Wohl, E.E. (2014). Multiscale structural and lithologic controls in the development of stream potholes on granite bedrock rivers. Geomorphology, 204: 588-598.

- Pelletier, J.D., Sweeney, K.E., Roering, J.J. and Finnegan, N.J. (2015). Controls on the geometry of potholes in bedrock channels. Geophysical Research Letters, 42: 7 pp.

- Sengupta, S. and Kale, V.S. (2011). Evaluation of the role of rock properties in the development of potholes: a case study of the Indrayani knickpoint, Maharashtra. Journal of Earth System Science, 120: 157-165.

- Springer, G.S, Tooth, S. and Wohl, E.E. (2006). Theoretical modeling of stream potholes based upon empirical observations from the Orange River, Republic of South Africa. Geomorphology, 82: 160-176.

- Udomsak, S., Choowong, N., Choowong, M. and Chutakositkanon, V. (2021). Thousands of potholes in the Mekong River and giant pedestal rock from north-eastern Thailand: introduction to a future geological heritage site. Geoheritage, 13: 17 pp.

- Whipple, K.X., Snyder, M.P. and Dollenmayer, K. (2000). Rates and processes of bedrock incision by the Upper Ukak River since the 1912 Novarupta ash flow in the Valley of Ten Thousand Smokes, Alaska. Geology, 28: 835-838.

Potholes and ecology

To our knowledge, specialist studies of potholes and ecology are less common, but one example based on investigations of the Wubu River near Chongqing City, China, is:

- Ren, H., Yuan, X., Yue, J., Wang, X. and Liu, H. (2016), Potholes of mountain rivers as biodiversity spots: structure and dynamics of the benthic invertebrate community. Polish Journal of Ecology, 64: 70-83.

General rivers books

Several books about the geomorphology and ecology of rivers have been published over the last few decades. Many are aimed primarily at university undergraduate students and tend to focus more on alluvial rivers rather than bedrock or mixed bedrock-alluvial rivers. Nonetheless, they contain useful overviews that may help those wishing to learn about rivers more generally. Examples include:

- Knighton, D. (1998). Fluvial Forms and Processes: A New Perspective (2nd edition). Hodder Arnold: London, 400 pp.

- Gordon, N.D., McMahon, T.A., Finlayson, B.L., Gippel, C.J. and Nathan, R.J. (2004). Stream Hydrology: An Introduction for Ecologists (2nd edition). John Wiley and Sons: Chichester, 448 pp.

Nature and ecology books

A number of books provide good introductions to general freshwater ecology, with some including field guides. 'Classic' and more recent examples include:

- Greenhalgh, M. and Ovenden, D. (2007). Freshwater Life: Britain and Northern Ireland (Collins Pocket Guide). HarperCollins: London, 256 pp.

- Hynes, H.B.N. (1970). Ecology of Running Waters. University of Toronto Press: Toronto, 569 pp.

- Giller, P.S. and Malmqvist, B. (1998). The Biology of Streams and Rivers (Biology of Habitats). Oxford University Press: Oxford, 296 pp.

Books about river landscapes in Wales

Many books provide fantastic photographs illustrating the stunning visual scenery of Wales but the geomorphological content – in other words, the explanation of the science behind the scenery – tends to be limited. An exception is the following book, which focuses on 100 of Wales's most remarkable landscapes, and blends geology and geomorphology with information from ecology, history, literature and other cultural connections. This includes several river locations with potholes and plunge pools, including Ffos Anoddun near Betws-y-Coed in Conwy, Pistyll Rhaeadr near Llanrhaeadr-ym-Mochnant in Powys, Sgwd Henrhyd-Nant Llech near Ystradgynlais in Powys, and Porth yr Ogof near Ystradfellte in Powys:

- Elis-Gruffydd, D. (2017). Wales: 100 Remarkable Vistas. Y Lolfa Cyf: Talybont, Aberystwyth, 311 pp.

Other general books about Welsh rivers include:

- Jones, J.L. (1986). The Waterfalls of Wales. Robert Hale Ltd: London, 242 pp.

- Clissold, P., Laws, T. and Sladden, C. (2012). The Welsh Rivers: The Complete Guidebook to Canoeing and Kayaking the Rivers of Wales (2nd edition). Chris Sladden Books: Westbury sub Mendip Wells, Somerset, 328 pp.

Other resources, including online

The internet provides a wealth of online resources about potholes, rivers and geomorphology in general, including many videos showing the land-shaping power of rivers in flood. Some resources specifically targeted at potholes and related features include:

Formation of potholes:
https://timeforgeography.co.uk/videos_list/rivers/formation-of-potholes/

Waterfalls, plunge pools and potholes:
https://www.bbc.co.uk/bitesize/clips/zqnqxnb

Waterfalls and gorges, erosion and deposition (follows the Afon Conwy):
https://www.bbc.co.uk/programmes/p00xptzz

A booklet (with supporting audio trail) for one of Wales's premier waterfalls and some spectacular potholes is:

- Tooth, S., Griffiths, H.M. and Llywelyn, S. (2017). The Natural Attractions of the Devil's Bridge Landscape: Answers to 10 Common Questions. Available at http://devilsbridgefalls.co.uk/nature-geomorphology/ [Welsh translation is: Tooth, S., Griffiths, H.M. and Llywelyn, S. (2017), Atyniadau Naturiol Tirwedd Pontarfynach: Atebion i 10 Cwestiwn Cyffredinol, also available at http://devilsbridgefalls.co.uk/nature-geomorphology/].

A guide to the natural and cultural aspects of the Ystwyth, Elan, Clywedog and Dyfi valleys of mid Wales, including suggested activities based around the periodically-exposed potholes and gorge at Pont Hyllfan on the Afon Elan, is:

- Tooth, S., Griffiths, H.M., Busfield, M., Llywelyn, S. and Thomas, A.D. (2018). Communicating Geoscience (Geomorphology and Quaternary Science): Guide to the Mid Wales Fieldtrip, Annual Meeting of the British Society for Geomorphology, Aberystwyth University, 10-14 Sept 2018, 57 pp.

A StoryMap tour of the Elan valley, including Pont Hyllfan, is:

- Llywelyn, S. (2019). Elan valley tour. https://www.elanvalley.org.uk/node/248903?language=en

The British Society for Geomorphology website has various resources, including an online colour booklet that outlines why geomorphology is important:

- Tooth, S. and Viles, H.A. (2014). 10 Reasons why Geomorphology is Important. Booklet produced on behalf of the British Society for Geomorphology. Available at: http://www.geomorphology.org.uk/publications/

A Welsh translation of the '10 Reasons' booklet has been prepared by:

- Griffiths, H.M. (2016). 10 Rheswm pam mae Geomorffoleg yn Bwysig. Available at https://www.geomorphology.org.uk/publications/

A version of the '10 Reasons' booklet specific to the Welsh landscape also has been produced:

- Tooth, S. and Griffiths, H.M. (2018). 10 Reasons Why the Geomorphology of Wales is Important. Booklet produced for the Annual Meeting of the British Society for Geomorphology, Aberystwyth University, 10-14 Sept 2018.

Aerial imagery and mapping

Many resources provide access to aerial imagery of river landscapes or to maps that will help place specific locations in context, including Welsh river reaches with potholes (see Section 7). Examples include:

- Google Earth: https://www.google.co.uk/intl/en_uk/earth/

- Google Maps (includes maps and satellite imagery): https://www.google.co.uk/maps/

- Streetmap (includes Ordnance Survey maps of the whole of the UK at different scales): https://www.streetmap.co.uk/

- Mapio Cymru: https://openstreetmap.cymru

- BGS Geology Viewer (a free app available on all browsers that lets you access detailed information about geology): https://www.bgs.ac.uk/map-viewers/bgs-geology-viewer/

SECTION 10
• SUGGESTED EDUCATIONAL ACTIVITIES •

Bedrock and mixed bedrock-alluvial river reaches with distinctive potholes and related features are excellent locations where activities that are both fun and educational can be undertaken.

As we have tried to demonstrate in this book, the various currents of river geomorphology, ecology, social history and culture interweave along river reaches with potholes (Sections 1 through 5), and spending time along such reaches can have significant health benefits while offering a range of informal learning opportunities (Section 6). More formal educational activities can be devised for all ages, but below we focus on those that might best be undertaken by teachers, parents and guardians of children, hopefully to inspire the next generation with a sense of wonder, respect and stewardship for landscapes. In schools, these activities need not be confined to bespoke lessons in geography, biology or history, but can cut across different lessons. Alternatively, they may simply be used as fun activities on a family outing. Most activities are best undertaken outdoors at river locations with potholes (see Section 7), but others could be undertaken indoors or online, perhaps before or after visiting such locations. Some activities can be supported by considering glossary terms (Section 8) and other resources, including publications, videos and animations (Section 9).

POTHOLE
A scoured depression in the bed or walls of a channel formed in cohesive substrate such as bedrock or clay; typically, deeper than wide and cylindrical in shape, but can have very complex shapes

Figure 10.1: Example of a pothole definition (from Wohl, E.E. (2013). Field and laboratory experiments in fluvial geomorphology. In: Shroder, J. (Editor in Chief) and Wohl, E.E. (Ed.), Treatise on Geomorphology. Academic Press, San Diego, CA, Vol. 9, Fluvial Geomorphology, pp.679-693).

OUTDOOR ACTIVITIES

- Observe, describe, sketch and photograph potholes. Describe their forms (plan view shapes) and textures. How many different types of potholes are there? How many conform to typical scientific definitions of potholes (see Figure 10.1); for example, in terms of a typical cylindrical shape?

- Measure pothole dimensions (widths/diameters, depths). What challenges do you face in undertaking these measurements? Do potholes tend to be deeper than wider, or wider than deeper?

- Look for evidence of past flow conditions. How high do floods get? Are there any potholes above the level of current floods, and if so, what do you think happens to these 'abandoned' potholes? How many potholes do you think formed at higher levels and have been lost as the river has eroded deeper into the rock?

- Describe the sediment within potholes. How big is the sediment: is it best described as clay, silt, sand, or gravel? For any gravel clasts, describe their forms (plan view shapes) and textures: do the clasts tend to be angular or round, and do they tend to be rough or smooth? How many different rock types are present among the gravel clasts?

- Describe and identify the rock type in which the potholes are eroded. How hard is the rock? Is it smooth or rough to the touch? Do potholes tend to be associated with joints, fractures and cracks? If so, how have these features affected potholes shapes and sizes? How do you think these potholes will continue to grow in the future?

- If making repeat visits to a river location with potholes, take repeat photos from the same perspective, especially after floods. Can you notice any changes after floods (e.g. in pothole size and shape, sediment volume and type, or ecology)?

- Observe and identify the plants growing in or near potholes (if necessary, use a reference book or smartphone app e.g. PictureThis). Why do you think those plants are found there? How have they adapted to the flow and sediment supply conditions?

- With care, observe and identify any fauna living in or near the potholes, particularly macroinvertebrates. How many different species are present? How do you think they survive during high flow (especially large floods) and low flows (especially droughts)?

- For small 'open' potholes without any sediment, flora or fauna, make casts (e.g. using clay).

- Use as many of your senses as possible (sight, sound, touch, smell, taste) to explore a single pothole or a series of potholes. While at the site, write a short description or poem (150 words or less) that captures your sensations.

INDOOR OR ONLINE ACTIVITIES

- Plot any field measurements of pothole dimensions (widths/diameters, depths) on graphs (e.g. line graphs, frequency histograms). Calculate some simple pothole statistics (e.g. mean widths and mean depths, width and depth ranges, width-depth ratios). Do the graphs and statistics confirm your field-based, visual impressions of whether potholes tend to be deeper than wider, or wider than deeper?

- Compare field measurements and graphs of pothole dimensions (widths/diameters, depths) from different locations on the same river, or locations on different rivers. What are the similarities and contrasts, and what might explain these similarities and contrasts?

- Use tools in Google Earth to characterise other aspects of locations with potholes (e.g. measurements of the width and height of waterfalls or the width and depth of gorges). What challenges do you face in undertaking these measurements? How do these measurements compare to the 'official' dimensions of waterfalls and gorges (e.g. as stated in tourist guides, official websites, or on-site signboards)?

- Compare your field photographs of potholes, waterfalls and gorges to historical photographs (e.g. see Pont Hyllfan potholes examples at: http://geoscenic.bgs.ac.uk/asset-bank/action/viewHome). Can you observe any differences in size or shape?

- Write definitions and produce a series of annotated diagrams to illustrate the processes of pothole, waterfall or gorge development.

- Plan, design and make a series of physical models (e.g. in clay) showing the development of a pothole, waterfall or gorge.

- Make a short video animation (2D or ideally 3D) that uses these definitions, diagrams and models to illustrate change over time.

- In view of these changes, when does a pothole stop being a pothole (e.g. potholes that coalesce in the process of gorge development)? Can we define an end point to the 'life cycle' of a pothole?

- Do you know of any alternative, local or regional names for potholes or related features? Can you invent any (e.g. 'pebble polisher')?

- Write some words and phrases linked to potholes and place them inside a physical or virtual model of a pothole. To facilitate this task, use a thesaurus to find synonyms for relevant words such as 'erode' and 'sculpt', and make use of devices such as alliteration (e.g. polished pothole) and onomatopoeia (e.g. 'burbling').

- Using typical definitions of potholes (Figure 10.1) or 'glossary' words relevant to potholes such as 'erosion', 'sediment' and 'vortex' (see Section 8), think of ways of writing or drawing them to show their meaning. For instance, could the shape of the words in Figure 10.1 be rearranged to represent different types of potholes (e.g. wide and shallow, narrow and deep) and/or could additional words be added or alternative words be substituted? With 'glossary' words, perhaps change the shape of words (e.g. into curves or circles; arrows could be used to indicate the direction of 'flow' of words). This could be done using pencil and paper, or using tools in Microsoft Word or equivalent software (Figure 10.2).

Figure 10.2: Example of word art that could be undertaken with words related to potholes (DR).

- Write a poem (e.g. an acrostic, haiku or kenning) about potholes, waterfalls or gorges, or other river features (Figures 10.3-10.5).

Powerful currents tumbling,
Over resistant rock.
Turbulent eddies swirl,
Hurtling pebble against pebble,
  and pebble against
Outcrop. Grinding, chipping, polishing, forming
Lasting cylindrical depressions.
Ecologically valuable, natural
Sculptural features.

Figure 10.3: Example of an acrostic for potholes (ST). In an acrostic, the first letter of each line provides the start of a word or phrase.

POTHOLE
Carved in bedrock
Round wonder of the river
Innate masterpiece

Figure 10.4: Example of a haiku for potholes (DR). A haiku is a form of Japanese poetry comprising 17 syllables, normally arranged 5-7-5.

Flow accelerator
Energy dissipator
Smoke maker
Noise factory
Water mixer
Pebble washer
Rock cleaver

Figure 10.5: Example of a kenning for a waterfall and plunge pool (ST). Kenning is a word from Old Icelandic and involves a multi-word substitution for a noun or nouns.

- Use presentation software (e.g. Microsoft PowerPoint or equivalent) to make an animated poem (i.e. moving text) or a poem in visual form.

- Plan, practise and refine a dance to mimic the geomorphological or ecological aspects of potholes (e.g. movement of pebbles, movement of migrating salmon).

- Write a reflective, personal account of a recent trip to a river location with potholes that invokes some or all of the different senses (sight, sound, touch, smell, taste), or produce a painting or drawing that captures some of these senses (e.g. Figure 10.6).

- Read and discuss travel writings, essays, poems or other accounts that refer to potholes, waterfalls and gorges. Use the examples with Welsh connections provided in this book and/or find other examples from Wales and farther afield (e.g. John Wesley Powell's accounts of his exploration of the Colorado River and its canyons in the southwest United States. See: http://www.gutenberg.org/ebooks/8082 and https://archive.org/details/explorationofcol1961powe)

- For students unable to visit river locations with potholes, use digital technology to help explain the key geomorphological, ecological, historical or cultural features of a given location or locations (e.g. using a virtual tour in Google Earth, or a short film that explains aspects of the location and makes use of some of the outputs and results from the above activities).

- Use digital technology to plan a real or imaginary tourist itinerary for different groups of people. For example: a week's holiday in Wales for a family of four (two adults, one young child, one older child) who are keen to see potholes, waterfalls and gorges (including gorge walking); a two week holiday in Wales for an older couple from the western United States who are adventurous and want to go 'off the beaten track' as well as visit some of the more popular tourist spots; a five day school trip with a focus on rivers to a country (or countries or region/s) of their choice.

- Using images, illustrate human impacts on river locations with potholes, waterfalls and gorges (e.g. pollution, flow control).

- For Pont Hyllfan on the Afon Elan or a similar location elsewhere, consider how the reservoir level dramatically impacts what features can be seen and when. To help with this task, perhaps use online resources such as National Library of Scotland maps where a historical map may be viewed next to a modern aerial image (see https://maps.nls.uk/geo/explore/side-by-side/). Use your case study as a stimulus for wider discussions about the pros and cons of dams and reservoirs in Wales and farther afield. How can we balance the societal benefits and negative social, geomorphological and ecological impacts of dams? Why are some dams promoted as tourist attractions? How do you feel about this marketing tactic?

Figure 10.6: Watercolour painting of Bourke's Luck potholes, South Africa (Graham Tooth).

ACKNOWLEDGEMENTS AND BIOGRAPHIES

Over many years, our investigations into the bedrock and mixed-bedrock alluvial rivers of Wales and farther afield have been supported by numerous organisations and individuals. Stephen and Hywel are grateful for funding and logistical support from organisations including: the Department of Geography and Earth Sciences (DGES) at Aberystwyth University, the University of the Witwatersrand in South Africa, the British Society for Geomorphology (BSG), the Coleg Cymraeg Cenedlaethol, the Joy Welch Educational Charitable Trust, and the Natural Environment Research Council (NERC). We thank Dr Janet Richardson and Dr Sioned Llywelyn for their postgraduate studies on aspects of Welsh bedrock river processes and geoheritage promotion, which have informed aspects of the content and approach of this book, and we thank colleagues in DGES and elsewhere for fruitful discussions. Tris Irvine-Fynn and Julian Ruddock kindly allowed us to include some of their photographs of rivers and potholes, and we acknowledge Antony Smith and Gareth Edwin for their assistance in the production of some of the figures. Any errors are, of course, our own. We are also very grateful to Myrddin ap Dafydd, designer Eleri Owen, and Gwasg Carreg Gwalch for their enthusiasm and care for the book. Finally, we thank our families for their support during our numerous fieldtrips to Welsh rivers and for acting as sounding boards for some of the ideas we discuss: Carys, Gwenno, Hamish, Maggie, Alaw, Lleucu, Morgan.

Dewi Roberts; image at bottom of page Lisa Barlow

BIOGRAPHIES

Dewi Roberts

Dewi Roberts has had a lifelong interest in rivers and is passionate about their geomorphology, wildlife and history. Dewi also revels in people's relationship to rivers creatively and emotionally and the stories and legends associated with them. He enjoys photographing and filming rivers, especially in the beautiful Welsh uplands and he has specialized in underwater photography; snorkeling helps him see deeper into the magical freshwater world. Known as 'The River Man', his footage has appeared on the BBC and S4C and he was featured recently on BBC's Springwatch. He explores rivers every day and in all weathers and is sometimes accompanied by his daughters. He likes seeing features such as potholes as natural sculptures and he has seen thousands upon thousands of them throughout Wales and farther afield. Dewi is particularly interested in how wildlife is adapted to life in running waters as well as the hydrodynamics involved in flow and the interplay of fluvial processes. Dewi considers himself to be immensely lucky to have such a strong relationship with rivers and knows first-hand of the huge benefits to health and wellbeing that can be gained from being in their presence.

Hywel Griffiths

Hywel Griffiths is a Reader in Physical Geography in the Department of Geography and Earth Sciences at Aberystwyth University, and his teaching and research interests focus on fluvial geomorphology, historical records and impacts of flooding and drought, science communication and creative geographies. He is also a poet and an author. His interest in rivers was inspired by exploring and playing on the banks of the small streams of rural Carmarthenshire where he grew up. He has been lucky to have been able to continue to explore rivers in Wales, Ireland, Spain, Crete, Patagonia and Jordan as part of his work and all these river bank journeys, whether abroad or close to home, inspire his creative work, even the short walk with the dog along the banks of the lower Afon Rheidol.

Stephen Tooth

Stephen Tooth is a Professor of Physical Geography in the Department of Geography and Earth Sciences at Aberystwyth University. His research and teaching interests mainly revolve around reconstructing past landscape changes, assessing rates of present-day landscape changes, and projecting future potential landscape changes resulting from climate variability and human activities. The main focus is on the role of rivers as agents of landscape change and he feels privileged to have travelled widely around the world looking at rivers in all their diversity. In recent years, he has been involved with activities promoting the role of the visual and non-visual arts in debates about global climate change and the Anthropocene, a proposed new interval of geological time that recognises humanity's impact on the functioning of the Earth system. These activities have included co-organisation of symposia, PhD supervision, art-science workshops, and public talks. He enjoys exploring the rivers and landscapes of Wales with family and friends, as this always provides fresh perspectives and new sources of inspiration.

Hywel Griffiths

Stephen Tooth

pp.130-131: Potholes on the Afon Dulas (north), near the border between Gwynedd and Powys (DR).